Multi-Scale
Discovery

Multi-Scale Approaches in Drug Discovery

From Empirical Knowledge to *In Silico* Experiments and Back

Edited by

Alejandro Speck-Planche

ELSEVIER

elsevier.com

Elsevier
Radarweg 29, PO Box 211, 1000 AE Amsterdam, Netherlands
The Boulevard, Langford Lane, Kidlington, Oxford OX5 1GB, United Kingdom
50 Hampshire Street, 5th Floor, Cambridge, MA 02139, United States

Library of Congress Cataloging-in-Publication Data
A catalog record for this book is available from the Library of Congress

British Library Cataloguing-in-Publication Data
A catalogue record for this book is available from the British Library

ISBN: 978-0-08-101129-4

Publisher: John Fedor
Acquisition Editor: Christine McElvenny
Editorial Project Manager: Anneka Hess
Production Project Manager: Mohanapriyan Rajendran
Designer: Vicky Pearson Esser

Typeset by TNQ Books and Journals

Contents

Contributors

M.F. Alves, Federal University of Paraíba, João Pessoa, Paraíba, Brazil

É.B.V.S. Cavalcanti, Federal University of Paraíba, João Pessoa, Paraíba, Brazil

R. Concu, University of Porto, Porto, Portugal

M.N.D.S. Cordeiro, University of Porto, Porto, Portugal

M. de Fátima Formiga Melo Diniz, Federal University of Paraíba, João Pessoa, Paraíba, Brazil

V. de Paulo Emerenciano, University of São Paulo, São Paulo, SP, Brazil

R. Dutt, G.D. Goenka University, Gurgaon, India

V. Garg, Maharshi Dayanand University, Rohtak, India

S. Golzio dos Santos, Federal University of Paraíba, João Pessoa, Paraíba, Brazil

H. González-Díaz, University of the Basque Country UPV/EHU, Bilbao, Bizkaia, Spain; IKERBASQUE, Basque Foundation for Science, Bilbao, Spain

V.V. Kleandrova, Moscow State University of Food Production, Moscow, Russia

G. Klebe, University of Marburg, Marburg, Germany

D. Leszczynska, Jackson State University, Jackson, MS, United States

J. Leszczynski, Jackson State University, Jackson, MS, United States

A.K. Madan, Pt. B.D. Sharma University of Health Sciences, Rohtak, India

L. Scotti, Federal University of Paraíba, João Pessoa, Paraíba, Brazil

M.T. Scotti, Federal University of Paraíba, João Pessoa, Paraíba, Brazil

A. Speck-Planche, University of Porto, Porto, Portugal

A.A. Toropov, IRCCS-Istituto di Ricerche Farmacologiche Mario Negri, Milano, Italy

A.P. Toropova, IRCCS-Istituto di Ricerche Farmacologiche Mario Negri, Milano, Italy

A.M. Veselinović, University of Niš, Niš, Serbia

J.B. Veselinović, University of Niš, Niš, Serbia

Chapter 1

Profiling Drug Binding by Thermodynamics: Key to Understanding

G. Klebe
University of Marburg, Marburg, Germany

1.1. THERMODYNAMICS: A CRITERION TO PROFILE PROTEIN–LIGAND BINDING

Lead optimization seeks for conclusive parameters beyond affinity to profile drug–receptor binding. One option is to use thermodynamic signatures since different targets require different binding mechanisms. Since thermodynamic properties are influenced by multiple factors such as interactions, desolvation, residual mobility, dynamics, protein adaptations, or changes in local hydration structure, careful analysis of why a particular signature is given can provide some insights into the binding event and help to define how a lead structure can be optimized.

In medicinal chemistry a given lead scaffold, possibly discovered by a fragment-based lead discovery campaign, is optimized from milli- to nanomolar binding (Wermuth, 2003; Blundell et al., 2002; de Kloe et al., 2009) either by "growing" the initially discovered scaffold into a binding site or by exchanging functional groups at its basic skeleton by other, purposefully selected bioisosteric groups. These modifications are intended to increase the binding affinity of the small-molecule ligand toward its protein receptor, and this usually results in an increase in the molecular mass of the candidate molecules to be improved.

To quantify this optimization process, ligand binding to its target protein is measured in terms of a binding constant (see in the following section), which is logarithmically related to the Gibbs binding free energy ΔG. This entity itself partitions into an enthalpic (ΔH) and an entropic ($T\Delta S$) binding contribution (Klebe, 2013; Chaires, 2008).

Since both properties ΔH and $T\Delta S$ contribute additively to the affinity ΔG, a desirable design strategy would optimize both properties in parallel. Is this, however, accomplishable at will, and of advantage in all cases? Without doubt,

Multi-Scale Approaches in Drug Discovery. http://dx.doi.org/10.1016/B978-0-08-101129-4.00001-1

different targets require ligands with different binding mechanism, which will possibly be mirrored by their thermodynamic profiles. A central nervous system drug needs different properties when compared with a drug addressing an extracellular target, e.g., a protease in the bloodstream. High target selectivity can be of utmost importance to avoid undesirable side effects; by contrast, promiscuous binding to several protein family members can be essential to completely downregulate a particular biochemical pathway, e.g., in case of kinases, or to achieve a well-balanced binding profile at a G-protein coupled receptor. Rapid mutational changes of viral or bacterial targets can create resistance against potent ligands. Pathogens follow this strategy by creating, e.g., steric mismatch or changes in the protein dynamics to diminish affinity of the bound active agent (Weber and Agniswamy, 2009; Ali et al., 2010). As the molecular foundations of these mechanisms are quite distinct, well-tailored thermodynamic signatures can be important to escape resistance. Freire et al. suggested improved resistance susceptibility for enthalpically optimized ligands exhibiting sufficient flexibility to evade geometrical modifications of the mutated target protein (Freire, 2008; Ohtaka and Freire, 2005). Equally well, ligands bind with entropic advantage (Fernandez et al., 2012) owing to high residual mobility, which allows the adaptation of multiple binding modes that may be beneficial to escape resistance development. This has been demonstrated by the superior resistance susceptibility of dapivirine or etravirine over other human immunodeficiency virus (HIV) reverse transcriptase inhibitors (Das et al., 2005). To apply such strategies in future projects we first need reliable access and interpretation of thermodynamic data. We therefore want to review in this contribution first the importance of different thermodynamic properties and how they can be determined by experimental methods. Furthermore, not only the accuracy and reliability but also the limitations of the experimental approaches to record thermodynamic data will be considered in the following. Subsequently, on the basis of these insights, we will discuss the thermodynamic signature across congeneric series of protein–ligand complexes to illustrate correlations between structure and thermodynamic profiles.

1.2. QUANTIFYING BINDING AFFINITY IN PROTEIN–LIGAND COMPLEX FORMATION

A binding constant can be determined either as a dissociation constant K_d, or inversely, as an association constant K_a under chemical equilibrium conditions (Ajay and Murcko, 1995). At a given concentration, it indicates which portion of a ligand is bound to the protein according to the underlying law of mass action. For enzymes, the so-called inhibition constant K_i is usually determined in a kinetic assay. The turnover of an appropriate substrate is recorded in a concentration-dependent manner. At low substrate concentration, it determines the dependence of the inhibitory concentration on the change in the turnover rate of the enzymatic reaction. Although K_i is not exactly defined as a dissociation constant, K_i, K_d, and K_a are usually referred to interchangeably and represent a kind of strength of the interaction between protein and ligand. Frequently, instead of the binding constant the so-called IC_{50} value is recorded. This value is characterized by the ligand concentration at which the protein activity has decreased to half of the initial amount. In contrast to K_i, IC_{50} values depend on the concentrations of enzyme and substrate used in the enzymatic reaction. The obtained values are affected by the affinity of the substrate for the enzyme, as substrate and inhibitor compete for the same binding site. Using the Cheng–Prusoff equation, IC_{50} values can, in principle, be transformed into binding constants (Cheng and Prusoff, 1973).

Under constant pressure and standard conditions (see later) the binding constant can be transformed into the Gibbs free energy of binding ΔG, which partitions into enthalpy ΔH and entropy ΔS, according to the equation $\Delta G = \Delta H - T\Delta S$. Spontaneously occurring processes are characterized by a negative free energy. At equilibrium, ΔG attains a minimum. The enthalpy reflects the energetic changes of interactions and desolvation associated with the various steps of protein–ligand complex formation. However, enthalpy changes are not the entire answer to why a complex is actually formed. In addition, it is important to consider changes in the ordering parameters. This involves how a particular amount of energy distributes over the multiple degrees of freedom of a given molecular system, composed of the ligand and protein prior to complex formation, the formed protein–ligand complex, and all changes that occur with water and the various components solvated in the water environment. Only if the system together with its surroundings transform into a less-ordered state, which corresponds to a situation of increased entropy, a particular process such as the formation of a protein–ligand complex will occur. Important enough the entropic component is weighted with temperature. It matters a great deal, whether the entropy of a system changes at low temperature, where all particles are largely in an ordered state, or whether it occurs at a high temperature at which the disorder is already significantly enhanced. Energetically favorable, exothermic processes are defined by a negative enthalpy contribution. If entropy increases, a positive contribution is recorded; however, because the entropic term $T\Delta S$ is considered with a negative sign, an increase in the entropy will cause a decrease in the Gibbs free energy and therefore an increase in binding affinity.

If binding affinity is discussed in terms of equilibrium thermodynamics, two aspects have to be kept in mind. Thermodynamics consider equilibria only; they do not tell anything about the kinetics, that is, how fast or whether even at all a particular equilibrium can be reached. Furthermore, biological processes occur, e.g., in the bloodstream or in a cellular environment where local concentrations constantly change with time. Such systems are only in first approximation in a steady-state situation of constant concentration. Finally, they have to be described as open systems, using nonequilibrium thermodynamics, which are by far more complex.

In principle, the thermodynamic properties are denoted with the superscript "°" to indicate that the values refer to standard states; however, this sign is often omitted. The necessity to refer to a standard state is to achieve comparability between measurements on a common scale for a mutual comparison (Krimmer and Klebe, 2015). Energies can only be measured as relative differences between two states, comparable with the determination of heights, e.g., of mountains. Usually, we refer the height of mountains relative to sea level to define a common and convenient reference point. In case of thermodynamic data, the binding free energy is referred to a standard state. This is defined as the conversion of 1 mol protein and 1 mol ligand to 1 mol of protein−ligand complex in a hypothetical ideal ("infinitely diluted") solution at constant pressure of $p^\circ = 10^5$ Pa. Such a solution has an activity coefficient of 1 ("activity" replaces "concentration" in real mixtures and the "activity coefficient" is a measure of the "effective concentration" of the species in a mixture, thus it describes the deviation from the originally weighted-in concentrations) (Pethica, 2015). The temperature is not part of the standard state and therefore has to be specified. The dissociation K_d and ΔH° are determined experimentally, e.g., in an isothermal titration calorimetry (ITC, see later) experiment, ΔG° is calculated using the relationship $\Delta G^\circ = RT \ln K_d$. The equation contains the natural logarithm of K_d, which is then used as a unitless value. Therefore, formally a standard concentration c° is used, which is set by convention as 1 M.

1.3. METHOD OF CHOICE TO ACCESS THERMODYNAMIC DATA: ISOTHERMAL TITRATION CALORIMETRY

The method of choice to obtain thermodynamic data is ITC (Chaires, 2008; Ladbury and Chowdhry, 1996; Ladbury, 2001; Velazquez-Campoy and Freire, 2005). ITC allows highly accurate determination of thermodynamic parameters without further requirement for chemical modifications such as labeling or immobilization. After an appropriate correction of superimposed effects such as the heat involving the exchange of protons with the surrounding buffer, the directly measured heat signal of an ITC experiment at a given temperature on titrating two compounds (e.g., protein and ligand) of known concentration provides the binding enthalpy ΔH and binding stoichiometry. From the shape of the titration curve the equilibrium binding constants (K_a or K_d) are determined and allow to directly calculate ΔG via $\Delta G = RT \ln K_d$. From an ITC experiment, ΔH and ΔG result simultaneously, and entropy is calculated as the numerical difference between ΔH and ΔG using the equation $\Delta G = \Delta H - T\Delta S$.

If protons are released either from the protein or the ligand or picked up from the buffer, a heat of ionization of the functional groups involved in the protein, ligand, and buffer substance(s) will be overlaid to the total heat signal (Jelesarov and Bossard, 1999; Baker and Murphy, 1996; Falconer and Collins, 2011). Whether such steps are involved can be determined performing the titration experiments at buffer conditions of varying heat of ionization, or with protein variants where a particular functional group has been exchanged (e.g., Tyr/Phe, Asp/Asn). Depending on the applied pH conditions, the molar ratio of entrapped or released protons is available from the Henderson–Hasselbalch equation. Quantifying the amount of protons exchanged between the involved functional groups requires knowledge of their pK_a values. This will also help to define where the protons actually go (Czodrowski et al., 2007). Such pK_a values can be measured, taken from tabulated data, or computed by programs. Yet, an important complication has to be regarded: pK_a values change with environment, thus substantial pK_a shifts may occur during complex formation, easily ranging over several orders of magnitude. As a result, some effort might be required, including mutagenesis of the likely involved residues, to trace which of the putative titratable groups of the system are actually responsible for the protonation (Neeb et al., 2014). This has to be taken into account while correcting for superimposed protonation effects. Even tricky cases can occur where protons are internally shifted between ligand and protein functional groups largely to the same amount (Baum et al., 2009a). As a result, no net protonation effect is observed, even though important changes do occur. Remarkably, heat-of-ionization effects are minor for oxygen functionalities and rather large for nitrogen-containing functional groups. The same holds for buffer compounds based on oxygen or nitrogen functional groups (Goldberg et al., 2002). Therefore, as a strategy to avoid at least major effects arising from the buffer, ITC experiments can be performed, e.g., in phosphate or acetate buffer where the buffer's heat of ionization is rather small; however, contributions from the functional groups of either the ligand or protein will still not be corrected. It is essential to correct for superimposed protonation steps in ITC experiments as heat effects will be superimposed to the actual binding signal. If remaining uncorrected, these heat effects will distort the enthalpic signal assigned to the ligand binding. As entropy is not measured but calculated from the numerical difference between ΔG and ΔH, a false partitioning of enthalpy and entropy will inevitably result. If such uncorrected data are used to interpret thermodynamic signatures, ill-defined correlations must be the consequence.

Overlaid protonation steps can also provide the chance to record thermodynamic data of entropic binders (Simunec, 2007). Binding of the latter ligands does not lead to any measurable heat signal. Only, if a protonation step is superimposed, the binding event may result with an exothermic or endothermic heat signal. Subsequently, the purely entropic binding profile only becomes apparent if the required correction is performed. Furthermore, it is essential to compare ITC experiments run at the same temperature. As ligand binding is predominantly related to a negative heat capacity change (see later), all binding signatures become enthalpically more favorable with increasing temperature (Jelesarov and Bossard, 1999). Accordingly, data collected at different temperatures can hardly be compared conclusively.

As mentioned, ITC experiments performed without control of possibly overlaid proton exchanges will be rather meaningless and false interpretation will likely result. If protonation changes occur within a congeneric series of protein–ligand complexes quite uniformly because, e.g., the protonation site is rather remote from the site where the congeneric ligands are actually modified, a comparison of the relative differences of the thermodynamic data is still justified and can be conclusively interpreted.

1.4. ISOTHERMAL TITRATION CALORIMETRY VERSUS VAN'T HOFF DATA TO ACCESS THERMODYNAMIC PROPERTIES

As mentioned, ITC experiments have the important advantage that two thermodynamic properties ΔH and ΔG result from one experiment, performed at the same temperature. Frequently, van't Hoff evaluations are performed to access thermodynamic parameters. In this case, the binding event is observed, usually via an easily recordable signal [e.g., photometric absorption, spectroscopic data, nuclear magnetic resonance (NMR), surface plasmon resonance (SPR), etc.] across a temperature range. To evaluate the measured data, the mostly rather inadequate assumption is made that the thermodynamic properties such as ΔH are temperature independent and may be determined by plotting in linear fashion the binding constants measured at different temperatures against the reciprocal temperature. For this, the so-called linear form of the integrated van't Hoff equation is used, which assumes ΔH to be temperature independent (Krimmer and Klebe, 2015). If the studied temperature range is rather small, an approximate linear correlation might be suggested and (inadequately) the slope of this linear correlation is extrapolated to assign a binding enthalpy. However, biological processes strongly depend on temperature; in consequence also the thermodynamic properties of these processes are temperature dependent. Thus, the van't Hoff equation cannot be straightforward integrated by assuming ΔH to be temperature independent over the considered temperature range. Instead, a nonlinear fit has to be at least applied (Liu and Sturtevant, 1995; Horn et al., 2001; Mizoue and Tellinghuisen, 2004). Furthermore, the van't Hoff evaluation assumes that the binding event follows a two-state transition between free and bound state and that the recorded signal change used to determine the binding constant reflects the entire population of free and bound molecules (Jelesarov and Bossard, 1999). As the correctness of this assumption is difficult to estimate, particular if the binding event passes through multiple states, the van't Hoff evaluation is even more difficult to justify. These considerations strongly argue to be very careful in using van't Hoff data as a source of thermodynamic binding information, at least when they are taken from a linear extrapolation. ITC appears, after appropriate corrections, as the more reliable information basis.

1.5. DETERMINATION OF HEAT CAPACITY CHANGES ΔCp

Another property, from a theoretical point of view a very informative one, is the change in heat capacity, ΔCp, at constant pressure of a biological system.

It indicates how well a system can absorb or release heat, thus it provides a crude idea how many degrees of freedom are available in the system to dissipate or store heat. Experimentally, ΔCp is available from ITC titrations performed at different temperatures (Jelesarov and Bossard, 1999). However, this evaluation and the subsequent interpretation run into similar complications as the van't Hoff evaluation. The considered multicomponent system as the formation of a protein–ligand complex is so complex, that even across a temperature range of 20–30K major structural changes will occur (e.g., already in the ubiquitously present bulk water phase) that make ΔCp interpretations extremely challenging. Consequently, it is usually rather problematic to discuss straightforward ΔCp changes of a protein–ligand complex system on molecular level, even though convincing examples have been reported (Stegmann et al., 2009).

1.6. THE ACCURACY AND RELEVANCE OF ISOTHERMAL TITRATION CALORIMETRY DATA

An important aspect addresses the accuracy of thermodynamic measurements (Tellinghuisen, 2012; Tellinghuisen and Chodera, 2011). Above all, the recorded data depend on the buffer composition and ionic strength of the ions used. Control experiments have been performed using the same biological system across different laboratories or different devices to estimate accuracy (Myszka et al., 2003; Baranauskienė et al., 2009). Purity of the ligands, stability of the proteins, constant water content, and avoidance of protein self-degradation or autoprotolysis have to be regarded.

Thus, how accurate can we expect ITC experiments to be? First, repetitive experiments have to be performed and averaged. Error propagation across interdependent properties has to be regarded. Besides calibration of the instrument, thorough control of the concentrations of the prepared solutions is important. Proteins are fragile compounds, and their activity depends on the way they were prepared, purified, and stored before usage. If protein solutions are prepared from solid material, the actual water content of the samples can be crucial. To achieve reliable results, it is highly advised to use material from the same batch for all experiments and to prepare solutions always freshly. Particularly proteases can decompose in concentrated solutions from autoprotolysis.

Usually, the ligand is titrated from a syringe with highly concentrated solution in a dropwise fashion into a large volume of the protein solution. In principle, this experimental setup can be reversed; however, limited solubility and availability or restricted stability of the proteins at high concentration impede the dropwise addition of a highly concentrated protein solution to the diluted ligand solution. Using the setup with the ligand added from the syringe, particularly the purity of high-affinity ligands is crucial for the accuracy of determining thermodynamic parameters. This results from the sigmoidal shape of the titration curve. For potent ligands, all injected molecules find in the beginning of the titration an unoccupied binding site. In due course of the experiment and after binding stoichiometry has been past, the heat signals reduce within a small amount of injections to the baseline when only the heat of dilution is still recorded. Uncertainties in the protein concentration will shift the binding isotherm leading to deviations in the expected stoichiometry, which are usually corrected for, assuming a 1:1 binding model. Thus, only minor deviations in the free energy determination are experienced. A much larger error will affect the determination of ΔH, which results from the integration over all heat signals. Ligand impurities can reduce these signals significantly and lead overall to smaller integrated ΔH values. As a result, an overestimated enthalpy/entropy compensation will be calculated. In case of weak binders, sigmoidal titration curves are hardly possible to record. Since fragment binding, particularly at high concentration, does not necessarily result in a stoichiometry of 1:1 (Mondal et al., 2014; Schiebel et al., 2016; Radeva et al., 2016a,b), the integration of the heat signals can become very inaccurate and such data are difficult to evaluate to reveal a reliable thermodynamic signature. Instead, displacement titrations can be used to make calorimetric analysis accessible for such ligands (Krimmer and Klebe, 2015; Zhang and Zhang, 1998; Rühmann et al., 2015a). They are also applicable to very strong binding ligands where the titration curve for the direct titration degenerates from sigmoidal to steplike shape making assignment of a K_d value unreliable (Valezques-Campoy and Freire, 2006).

Considering all these factors properly including the correction for superimposed protonation events, an evaluation across a series of congeneric compounds in terms of relative differences will cancel out most of the systematic errors. In favorable cases, the accuracy can amount to about 1 kJ/mol, particularly if relative comparisons of two related ligands are performed (Krimmer and Klebe, 2015).

The ITC experiment records all changes involving heat effects going from the individually solvated protein and ligand to the newly formed protein−ligand complex. Besides conformational and configurational changes of the binding partners, protein and ligand, this process involves also substantial changes in the water structure. However, the binding event is a multistep process; all modifications are finally compressed into the three thermodynamic parameters ΔG, ΔH, and −TΔS, and they represent the entire complex formation process. Subsequently, we are tempted to relate the changes of these parameters with the binding event and solely focus rather naively on the newly formed protein−ligand interface. However, much more is involved that might reflect changes in the protein structure, e.g., activation/deactivation of conformational, vibrational, or rotational degrees of freedom of protein side chains remote from the binding site or rearrangements of the water structure across the surface of the involved components, i.e., ligand, uncomplexed protein, and newly formed protein−ligand complex. Even compensating entropy−entropy effects have been reported involving locally deviating activation and attenuation of rotational degrees of freedom of methyl group side chains (Homans, 2005; Kasinath et al., 2013). All these contributions will have an impact on the thermodynamic signature of the binding event. As will be shown in the following sections of this review, the presence or absence of a single water molecule next to the protein–ligand interface can easily shift the thermodynamic profile in enthalpy and entropy by 5–7 kJ/mol in either direction. Usually, within a series of congeneric ligands, we tend to interpret effects of this magnitude as significant and they might give rise to a contrary interpretation of the binding event, even though, unexpectedly, only the difference of one involved water molecule is responsible for the distinct thermodynamic profile. This can easily lead to misinterpretation, particularly if, rather superficially, a particular drug candidate is assessed as superior, e.g., for its more enthalpic profile (Ladbury et al., 2010). To reduce the danger of misconception, the consideration of complementary information is of utmost importance. In consequence, interpretation of thermodynamic data, even across a very narrow congeneric ligand series, will hardly be meaningful without monitoring the structural properties of every formed protein–ligand complex simultaneously. Such information is available from high-resolution crystallography, and the concomitant survey of the produced complexes by crystal structure analyses (or/and NMR) is an inevitable requirement for the meaningful interpretation of thermodynamic signatures.

Even in such ideal cases where the corresponding crystal structures are available, some caveat is given. ITC data are recorded at ambient temperature in a buffered solution. Structural data, however, are collected in the crystalline phase often enough at liquid nitrogen temperature. Thus, can any correlation between solution and crystalline state be expected? Recent comparative diffraction studies performed at ambient and low temperature revealed differences in the scatter of side chain torsion angles (Fraser et al., 2011; Fenwicka et al., 2014). Supposedly, these molecular degrees of freedom are soft enough to still allow motion and adjustment under the flash cooling protocol applied to freeze protein crystals for diffraction experiments. They will adjust with temperature. Other motions involving larger rearrangements cannot occur in the crystalline phase, for example the complete rearrangements of water surface layers. Here, flash-cooled crystals will likely mirror the situation at ambient temperature. In several of our investigated compound series, we observed a qualitative correlation of the B-factors, which are attributed to residual thermal motion in a crystal with entropic effects monitored by ITC in solution (Baum et al., 2010; Baum et al., 2009b; Neeb et al., 2016). Therefore, at least qualitatively, a correlation between ITC and crystal structure data seems to be given allowing for a conclusive discussion of structures along with thermodynamics. This estimation matches well with the conclusions of Nakasako (2004), who compared solvation patterns of water molecules observed in crystal structures under cryo conditions with other physicochemical measurements and found high consistency. These findings make us confident that crystal structures are actually relevant for the interpretation of ITC data.

1.7. PROTEIN–LIGAND COMPLEX FORMATION: WHAT CAN THERMODYNAMIC DATA TELL ABOUT A GOOD STARTING POINT FOR OPTIMIZATION?

Subsequent to these general considerations about the access and accuracy of thermodynamic data, we want to discuss possible strategies how to make use of thermodynamic signatures with respect to lead optimization. As discussed previously, an ITC experiment gives access to the simultaneous determination of ΔH and ΔG. Using the fundamental equation $\Delta G = \Delta H - T\Delta S$, the entropic contribution is calculated from the numerical difference between ΔH and ΔG.

What appears on first sight as a very convenient access to enthalpy and entropy bears an important caveat. Since both entities do not result from independent experiments, but from a numerical difference, they are inevitably correlated and will automatically compensate to match with the measured free energy. This means that any undetected systematic error or uncorrected effect in the measured enthalpy will automatically be reflected in a compensatory change of the entropic contribution (Krimmer and Klebe, 2015; Sharp, 2001; Olsson et al., 2011; Chodera and Mobley, 2013). Nonetheless, enthalpy/entropy compensation also occurs as an intrinsic physical phenomenon (see later) (Dunitz, 2003), but it has to be kept in mind that experimental deficiencies, inappropriate data corrections, or reference to inadequately defined standard states required to perform global comparisons will give rise to some inevitable, from a physical point of view, irrelevant enthalpy/entropy compensations.

Prior to complex formation, both, the protein and ligand, are separately solvated and move freely in the bulk solvent phase. Upon complex formation, the two independent particles merge into one species. On achieving this, they sacrifice rotational and translational degrees of freedom as two independent particles reduce to one (Murray and Verdonk, 2002). This entropic loss was calculated to amount to about 16 kJ/mol and is associated with a price in Gibbs free energy. Experimentally, this value is nicely confirmed by Nazare et al. (2012) and Borsi et al. (2010), who studied the merging of two nonoverlapping fragments binding to factor Xa or matrix metalloproteinase-12. Comparing the affinity of the two individual fragments with that of the merged "supermolecule" reveals values of 14–15 kJ/mol, which match well with the theoretically determined entropic cost for the loss of degrees of freedom for merging two particles into one.

This also sets a lower affinity limit for complex formation to about −15 kJ/mol, as reflected in a thermodynamic data compilation of Olsson et al. (2008). The authors have collected published ITC data and plotted the information on $\Delta H/-T\Delta S$ diagram (Fig. 1.1). The main diagonal in this plot corresponds to the observed data scatter in Gibbs free energy, which covers a range from approx. −15 to −60 kJ/mol. This distribution reflects the range accessible to medicinal chemists for ligand optimization from millimolar to subnanomolar

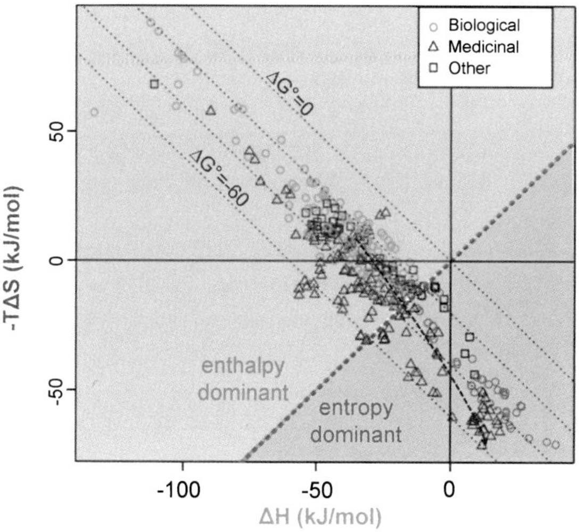

FIGURE 1.1 Published isothermal titration calorimetry data plotted onto a ΔH/ TΔS diagram. The diagram has been split into areas where enthalpy (green) or entropy (red) dominates the Gibbs free energy of binding (ΔG). Along the main diagonal the scatter in ΔG from about −15 to −60 kJ/mol is found corresponding to the range accessible to medicinal chemistry. Perpendicularly, the mutual scatter in enthalpy and entropy with opposing contributions to ΔG is shown. It spreads across a very large range indicating intrinsic enthalpy/entropy compensation. Data are classified for ligands coming from different sources; optimization resulting from medicinal chemistry programs tend to improve for entropic reasons (*blue dashed arrow*). *The figure was taken with permission from Olsson, T.S.G., Williams, M.A., Pitt, W.R., Ladbury, J.E., 2008. The thermodynamics of proteinligandinteractions and solvation: insights for ligand design. J. Mol. Biol. 384, 1002−1017 with slight modifications.*

affinity. The secondary diagonal, perpendicular to the ΔG distribution, displays the mutual scatter in enthalpy and entropy with opposing contributions to ΔG. As this distribution spreads over a very large range, it reveals an inherent enthalpy/entropy compensation, leading to the rather small scatter in ΔG.

The enthalpy/entropy diagram can be dissected into one area where enthalpic (green) and another where entropic contributions (red) dominate. Remarkably, when compared with biomolecules, ligands originating from medicinal chemistry optimization tend to display in this diagram an enhanced entropic binding profile with growing potency (blue arrow). This observation has provoked in the past the question as to whether a more enthalpically or entropically driven binding is desired (Freire, 2008; Ladbury et al., 2010) and whether such a ligand binding profile can be designed at will (Freire, 2009)? The intrinsic enthalpy/entropy compensation suggests that both properties are interdependent in reciprocal manner, thus can they be optimized independently or both in parallel? The latter would optimize ΔH and −TΔS simultaneously to achieve the most efficient ΔG enhancement, but is such a strategy achievable in light of the enthalpy/entropy compensation? Even though there is no physical law arguing for mutual enthalpy/entropy compensation, considerations on molecular level suggest that both opponents will at least partially cancel out (Dunitz, 2003). Summarily, strong enthalpic interactions will fix a ligand at the binding site, which is entropically unfavorable. By contrast, pronounced residual mobility in the bound state is entropically beneficial, as a smaller number of degrees of freedom is lost upon complex formation. However, the quality of the formed interactions will be less efficient leading to a minor enthalpic contribution.

The correlation diagram shown in Fig. 1.1 [and similar evaluations that have been published in literature (Olsson et al., 2008; Hann and Keserü, 2011; Ferenczy and Keserü, 2010, 2016; Reynolds and Holloway, 2011)] implies that the published ITC data would have all been properly corrected for superimposed protonation changes and refer to a common standard state. Supposedly, this is in most cases not given making any global comparison of such data quite questionable and the conclusions drawn can be easily misleading (Krimmer and Klebe, 2015). The aforementioned comparison with the height of mountains should be consulted again to explain this issue. As mentioned, the height of mountains is conveniently referred to sea level; however, also other scales could be imagined, e.g., the midpoint of the earth, the lowest depression on a continent, or the deepest point found in any of the oceans. Many such reference points are imaginable. However, if everyone would decide to use a different reference point to measure heights, it would be impossible to compare mountain heights globally on a comparative map. For a mountain climber who plans to ascent a next summit and to estimate on the required resources, it is only important to know the relative height of the surrounding summits with respect to the point from where he starts his trip. Microcalorimetric measurements are setup under very different conditions, mostly optimized for the system studied (added salts, buffers, detergents, added cofactors or co-substrates, DMSO, etc.) but hardly adhering to standard conditions, particularly with respect to concentrations and the assumed "ideal solution" conditions (see earlier). This makes global comparison of such data nearly impossible, and falsely inevitable enthalpy/entropy compensations will affect the data to be correlated. We studied some systems under varying salt, DMSO, or detergent concentrations with and without added cosubstrate and observed that the Gibbs free energy is only little affected; however, the measured heat signal changes fairly strongly, thus affecting the partitioning of the derived ΔH and $-T\Delta S$ values on absolute scale. Nevertheless, what matters in this case is the important observation that within a congeneric series of ligands the *relative* differences in the changes of enthalpy and entropy ($\Delta\Delta H$ and $-T\Delta\Delta S$) remain virtually unaffected by the applied conditions. This underscores that the comparison of thermodynamic data taken from ITC measurements should only be performed across congeneric series on *a relative scale*. Considering our mountain climber, his decision regarding which summit to climb with the resources available to him can be taken considering the relative difference of the summit heights whereas the absolute heights are not important for this decision.

1.8. OPTIMIZATION: GO FOR ENTHALPY OR ENTROPY?

Some aspects with respect to ligand optimization should be considered (Hann and Keserü, 2011; Ferenczy and Keserü, 2010, 2012, 2016; Reynolds and Holloway, 2011; Martin and Clements, 2013). In the beginning of a hit-to-lead optimization, particularly hydrogen bonds are added, often under the guidance of structure-based drug design. This early phase of drug optimization is predominantly driven by enthalpic considerations as we will see in the following section. However, during late stage optimization proceeding from low micro- to nanomolar range, drug candidates are often optimized by locking them in the receptor-bound conformation using a rigidified ligand scaffold (Rühmann et al., 2016; Glas et al., 2014). This preorganization of the ligand in its bound conformation reduces the conformational multiplicity in the unbound state, and affinity is increased primarily for entropic reasons (Fig. 1.2) (Rühmann et al., 2016). Other strategies follow the attachment of lipophilic groups to increasingly fill hydrophobic protein pockets (Martin and Clements, 2013; Biela et al., 2012a), which is attributed to the classical hydrophobic effect (Kyte, 2003, Chandler, 2005; Dill et al., 2005). If strategies focused on entropic optimization are pursued at late stage optimization, it appears advisable to start with an enthalpically binding hit as the entropic component will be subsequently added anyhow (Ladbury et al., 2010; Olsson et al., 2008). A prerequisite to follow this strategy is a reliable characterization of the initial hit selected for the optimization of the binding signature. In Fig. 1.3, two series of thrombin inhibitors are shown, which were optimized by sequentially attaching identical substituents (Biela et al., 2012a). The relative differences in the thermodynamic profiles are virtually the same, but the

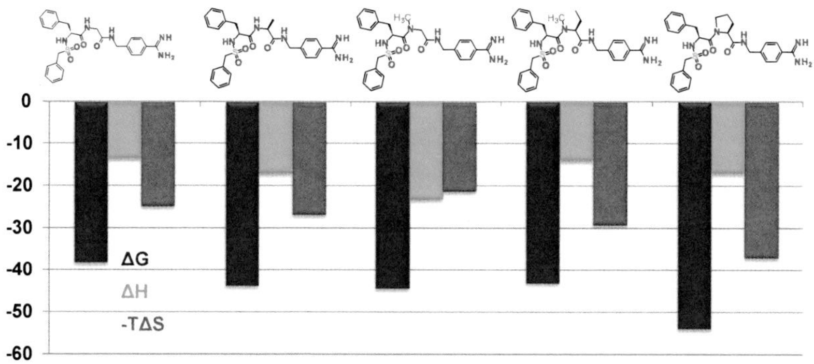

FIGURE 1.2 Difference of the relative thermodynamic binding data (ΔΔG, ΔΔH, −TΔΔS) of five ligands (left to right) optimized by preorganizing their geometry in the protein-bound conformation. In this figure as in all following ones the thermodynamic data, recorded at 25°C, are shown in terms of histograms. The example shows that the glycine derivative (left) inhibits thrombin much more potent than the proline analog (most right) mainly for entropic reasons.

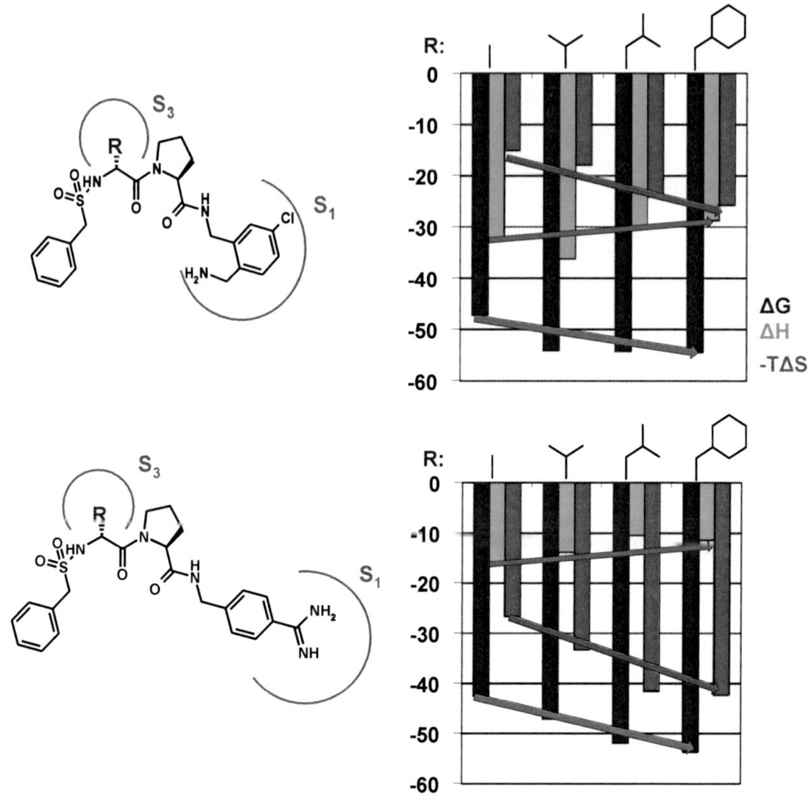

FIGURE 1.3 Two congeneric series of thrombin ligands optimized to fill the S_3 pocket with size-increasing hydrophobic substituents (R). Affinity (ΔG) improves similarly in both series due to stronger entropic gain than enthalpic loss. Nonetheless, the *m*-Cl-benzyl series displays overall the more enthalpic binders compared to the benzamidine series.

benzamidine series starts from a more entropically favored level than the *m*-Cl-benzyl series. In consequence, the former series would be classified as more entropically, and the latter as more enthalpically driven. It appears rather arbitrary; however, to use this difference as a criterion to rank the second series as "more promising" for a drug development program, simply for its "enthalpic advantage." Considering the *relative differences* (regard only the slope of the indicated arrows in Fig. 1.3) among the members of both series comes to a conclusive result: The affinity enhancement is achieved by an entropic advantage across the series, which is to some part compensated by an enthalpic loss.

Particularly for weak binding ligands, which are often used as starting point for lead optimization, this thermodynamic profiling is, as mentioned earlier, rather difficult. This occurs for the following reason. Two structurally very similar and nearly equipotent aldose reductase inhibitors, shown in

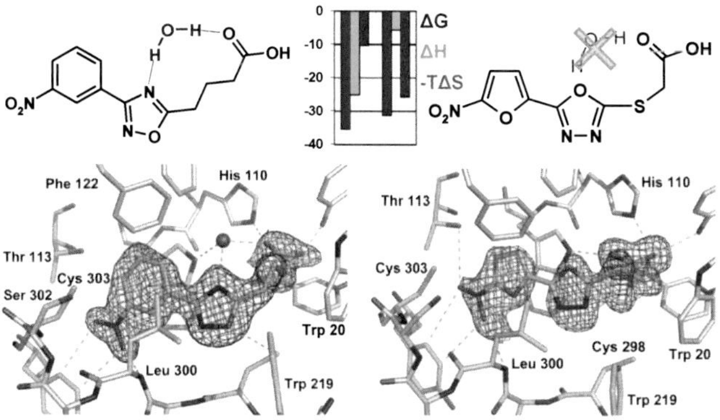

FIGURE 1.4 Two structurally related ligands bind to aldose reductase with similar affinity and nearly identical binding mode. Nonetheless, ΔG partitions differently into ΔH and −TΔS which can be explained by the entrapping of a water molecule mediating a contact between the enthalpically favored ligand and the protein (left).

Fig. 1.4, differ by being either a more enthalpic or entropic binder (Steuber et al., 2007a). The enthalpically more favored ligand captures a water molecule mediating a protein−ligand interaction, whereas the entropically favored one lacks this interstitial water molecule. In the latter case, the entropic cost for entrapping the water molecule has not to be afforded; consequently, the overall profile appears with an entropic advantage.

In modern drug discovery, many development programs start with fragments exhibiting a molecular weight between 150−250 Da (Blundell et al., 2002; de Kloe et al., 2009; Rees et al., 2004; Erlanson et al., 2004). Even though being weak binders, they possess remarkable binding affinity with respect to the actual size. This is conveniently expressed by their high ligand efficiency (Hopkins et al., 2004). To select the most promising fragment for subsequent optimization, it would be highly desirable to characterize the thermodynamic binding signature of fragments (Ferenczy and Keserü, 2012, 2013; Keserü et al., 2016). Unfortunately, they often show a faint heat signal of binding, usually too weak to allow for direct ITC titrations. In consequence, the absorbed or released heat signal is too low to produce a sigmoidal titration curve even at very high concentrations. In consequence, it is not possible to observe the inflection point of the titration curve and the stoichiometry has to be calibrated (Krimmer and Klebe, 2015; Rühmann et al., 2015a). Sophisticated displacement titration protocols can overcome this problem, however, for the sake of a reduced accuracy of the determinations owing to error propagation of multiple titration experiments (Krimmer and Klebe, 2015; Rühmann et al., 2015a; Zhang and Zhang, 1999; Valezques-Campoy and Freire, 2006). In Fig. 1.5, the binding profiles of 66 fragments binding to endothiapepsin are shown. As with larger ligands, fragments also scatter over the entire range from an enthalpically to entropically more favored binding

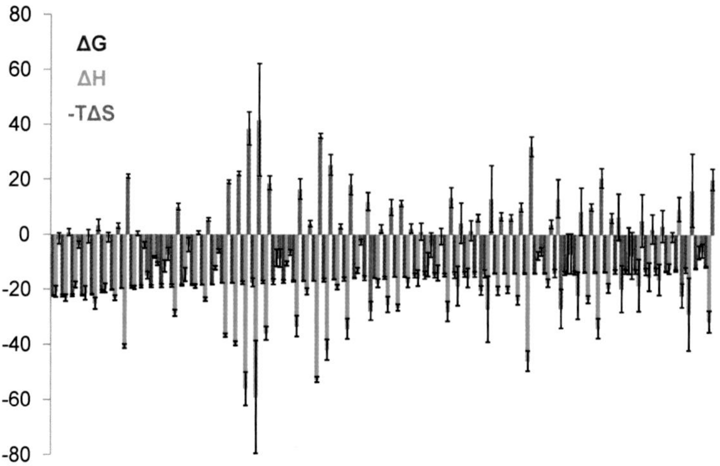

FIGURE 1.5 Thermodynamic signature for 66 fragments binding to the aspartyl protease endothiapepsin. The fragments are sorted from left to right by decreasing ΔG, ΔH, and −TΔS scatter over a large range. For some of the complexes the profile could only be determined with reduced experimental accuracy (cf. error bars in black).

(Schiebel et al., 2016; Huschmann et al., 2016). As fragments fill the binding pocket only partly, their binding is strongly influenced by the release, shift, or uptake of water molecules. Since water molecules can significantly modulate the binding profile, structural characterization of the water pattern in fragment binding is a prerequisite to allocate and decipher the thermodynamic signature of the binding properties of a fragment. In Fig. 1.6, three fragments binding to thrombin are compared (Rühmann et al., 2015b). The enthalpically most favored chlorothiophen fragment displaces one water molecule more from the protein-binding pocket compared with the other two amidino-type fragments,

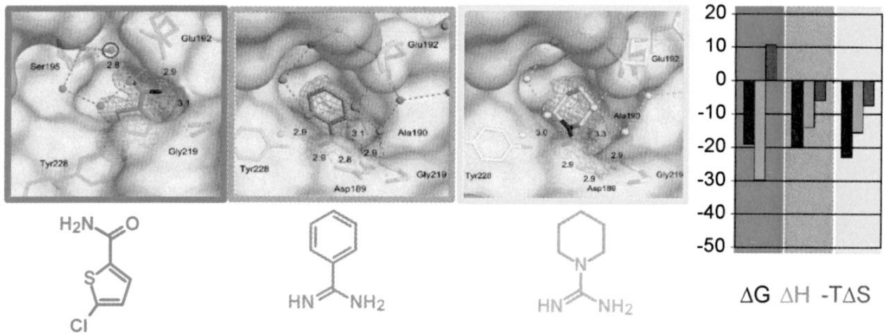

FIGURE 1.6 Thermodynamic signature for three fragments binding to thrombin. In the upper row, the high-resolution crystal structures with the enzyme are shown. In the lower row, the thermodynamic data are shown rendering the chlorothiophen fragment as most enthalpic binder. It displaced, when compared with the two amidino derivatives, one water molecule more from the S_1 pocket of the protease.

which exhibit very similar profiles. The displacement of this water molecule (see later) results in a strong enthalpic advantage, which is remarkably reflected in the thermodynamic signature of the fragment.

1.9. WHAT DOES AN H-BOND OR A LIPOPHILIC CONTACT CONTRIBUTE?

The thermodynamic profile of a complex will change, if a functional group, competent to establish hydrogen bonds, is introduced; however, by how much and how does the ΔG enhancement partition in enthalpy and entropy? The strength and consequently the affinity contribution of a hydrogen bond strongly depends on the charges carried by the mutually interacting functional groups. Upon mutation of the native Leu in the wild type of aldose reductase for Pro, we could replace the NH group forming an uncharged H-bond to fidarestat (Fig. 1.7A). The Gibbs free energy difference for the loss of this "neutral" H-bond accounts for $\Delta\Delta G = -7.8$ kJ/mol. The partitioning shows that it is mainly of enthalpic nature. If in the wild type the residue Tyr48 is replaced by Phe, a charge-assisted H-bond to the inhibitor IDD594 is lost (Fig. 1.7C) (Steuber et al., 2007b). The binding mode is conserved, and the H-bond relates to a $\Delta\Delta G = -8.5$ kJ/mol difference. Interestingly in the latter case, the enthalpic portion $\Delta\Delta H$ contributes -22.7 kJ/mol, which is significantly compensated by an unfavorable entropic portion of $-T\Delta\Delta S = +14.2$ kJ/mol. In a series of thrombin inhibitors, the establishment of a charge-assisted H-bond comes to a very similar enthalpic contribution (Baum et al., 2010); however, as the overall affinity enhancement varies with the attached substituents, an entropic signal is imposed that largely diminishes the enthalpic effect (Fig. 1.7D). Finally, a salt bridge formed between the ligand's amidino group and Asp189 of thrombin results, when compared with the unsubstituted analog, in a $\Delta\Delta G$ gain of -14.4 kJ/mol (Fig. 1.7E) (Baum et al., 2009b). This involves a huge enthalpic component of $\Delta\Delta H = -26.5$ kJ/mol, partly compensated by an unfavorable entropic penalty of $-T\Delta\Delta S = +12.0$ kJ/mol. Overall, this compilation shows that additional hydrogen bonds enhance binding mainly for enthalpic reasons, but a significant entropic contribution reduces the overall affinity improvement, particularly if charges are involved.

Another strategy to enhance binding focuses on the burial of ligand hydrophobic portions in hydrophobic protein pockets. In a congeneric series of six HIV protease inhibitors we exchanged the terminal substituent and increased its hydrophobic surface resulting in unchanged binding poses (Fig. 1.8A). If the increase in the molecular volume is plotted against affinity, a linear relationship is obtained. From the slope of the correlation, a surface patch increment of -65 J/molÅ^2 to affinity can be deduced (Blum, 2007). This increment matches well with the range found by the statistical evaluation of a large sample of protein–ligand complexes (Olsson et al., 2008). Nonetheless, the increments scatter over a large area.

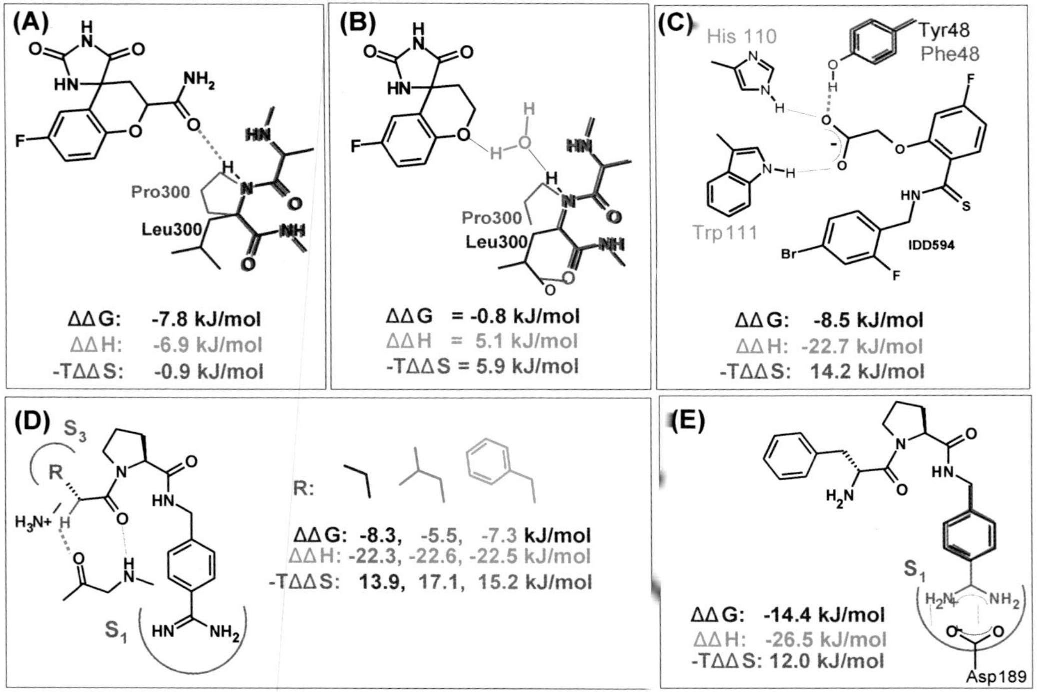

FIGURE 1.7 Thermodynamic contributions of hydrogen bonds to protein–ligand complex formation. (A) An uncharged H-bond is formed between fidarestat and aldose reductase (blue) whereas this bond is missing in the Pro variant (red). The affinity improves mainly for enthalpic reasons. (B) The analog Sorbinil binds to aldose reductase mediated through an interstitial water molecule. Upon Leu300Pro replacement the water molecule and the mediating H-bond contacts are lost. Overall, this loss is hardly reflected in affinity, however; strong compensating effects in enthalpy and entropy are experienced. (C) If charges are involved the enthalpic contributions of a hydrogen bond increase strongly, e.g., if in the wild type Tyr48 is replaced by Phe, or (D) in thrombin if the ligand's charged ammonium group is replaced by a hydrogen incapable to form an H-bond. However, the enthalpic contributions are very similar, and the affinity enhancements vary due to a detrimental entropic signal of deviating amount. (E) If a salt bridge to Asp189 in thrombin is introduced, an even larger enthalpic contribution is experienced, partly compensated by an adverse entropic penalty.

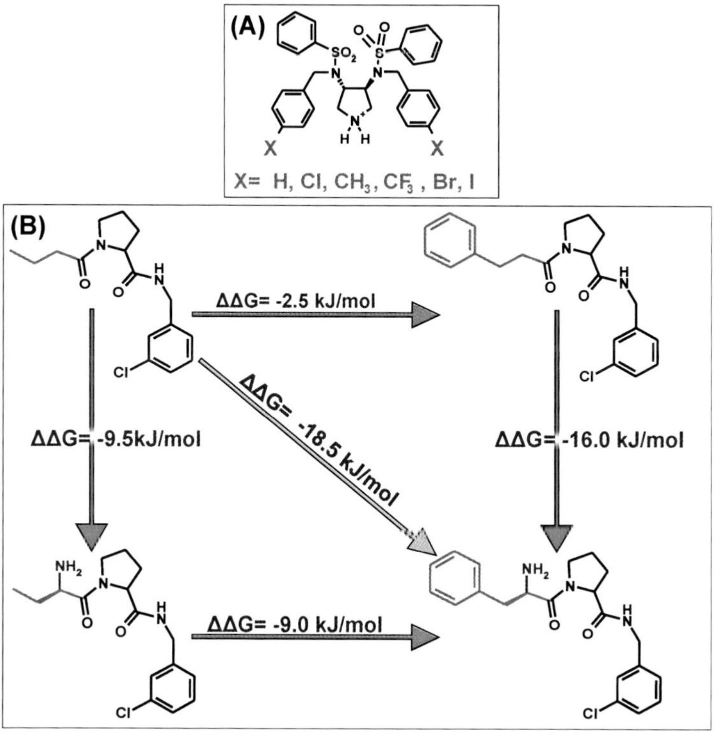

FIGURE 1.8 (A) Addition of hydrophobic substituents at human immunodeficiency virus protease inhibitors increases protein binding by lipophilic surface-patch contributions of -65 J/molÅ^2. (B) Inhibitor binding to thrombin has been studied for a series of systematically varied ligands. Attachment of a lipophilic side chain and an H-bond forming group have been added with deviating sequence. Adding hydrophobicity first and then the H-bonding group results in different affinity increments compared to the strategy where the H-bonding anchor is attached first followed by the lipophilic group. This phenomenon of cooperativity finds an explanation in the deviating residual mobility of the attached lipophilic side chain depending on the presence or absence of the H-bond.

Yet, which thermodynamic property drives the affinity enhancement in this case? The attachment of increasingly lipophilic substituents is usually followed at a late-stage optimization when ligand binding has passed from micro- to nanomolar potency and the binding mode of the lead scaffold is well characterized. At this stage, ligands are conceptually modified to rigidify the bound conformation (see earlier example, Fig. 1.2). The series of thrombin inhibitors, which place a P_3 substituent of growing lipophilicity into the S_3 pocket of the protease, have already been introduced in Fig. 1.3 (Biela et al., 2012a). As P_1 anchor, one series featured a charged benzamidine moiety, and the other a 2-(aminomethyl)-5-chlorobenzylamide anchor. The gradual increase from methyl to cyclohexylmethyl at P_3 enhances binding mainly for entropic reasons. On the structural level, a successive replacement of water molecules, found in well-ordered positions prior to ligand binding, is experienced.

The increase in lipophilicity of ligands results in an affinity increase predominantly for entropic reasons, particularly if previously ordered water molecules are released from the binding pocket to the bulk water phase (Biela et al., 2012a). Nonetheless, enthalpic signatures have also been reported for hydrophobic binding. In such cases, the replacement of the nonoptimally H-bonded water molecules (the so-called unhappy waters with respect to the bulk water phase) or the filling of suboptimal hydrated binding pockets is held to be responsible for the unexpected profile (Homans, 2007; Englert et al., 2010; Snyder et al., 2011; Wang et al., 2011; Setny et al., 2010).

1.10. PAIN IN THE NECK: H-BONDS AND LIPOPHILIC CONTACTS ARE MUTUALLY DEPENDENT

If enhancement of ligand binding by additional hydrogen bonds and burial of growing lipophilic surface patches are optimization strategies, mirrored by deviating thermodynamic profiles, the question arises whether they are mutually interdependent and whether their contributions tend to mutually compensate?

In a study with thrombin ligands, we increased the hydrophobic surface of the P_3 substituents to grow into the S_3 pocket of thrombin. Modeling studies suggested a surface patch increment of -78 J/molÅ^2 (Muley et al., 2010). Adding an adjacent amino group to the ligand scaffold, competent to establish a charge-assisted H-bond with the protein, this patch increment augments to -128 J/molÅ^2. Obviously, a significantly larger value is suggested when compared with the series lacking the amino group. From a rational point of view, this observation appears quite paradoxical.

Detailed analysis unravels that the introduction of the additional hydrogen bond and the increase of the hydrophobic substituent are not independent of each other (Fig. 1.8B). Optimization can either start by increasing the size of the hydrophobic substituent and subsequently adding the amino group, or conversely, first introducing the amino group and then the lipophilic substituent. As the affinity increase along the two alternative pathways shows, strict additivity of the functional group contributions does not apply: instead, the H-bond and growing lipophilic contact exhibit mutual cooperativity. Molecular dynamics simulations of the involved complexes indicate that the observed cooperativity results from distinct dynamic properties of the complexes. If the hydrogen bond is absent and the hydrophobic substituent is of small size, enhanced residual mobility of this moiety in the S_3 pocket of the protein is observed. Averaged over time, a less efficient surface contact can be observed between the hydrophobic ligand and the protein. Introduction of the H-bond as well as the exchange from small- to medium- and large-sized hydrophobic substituent in S_3 results in reduced residual mobility of the attached substituent. In consequence, averaged over time a more efficient hydrophobic contact is experienced between ligand and protein. This enhancement owing to the mobility restricting H-bond matters more in case of the smaller substituents. By contrast, for the larger hydrophobic substituents residual mobility is anyhow limited owing to more efficient space filling of the S_3 pocket. Therefore, with and without the presence of the hydrogen bond a more efficient lipophilic contact is experienced on the time scale. Thus, the observed cooperativity finds its explanation in changes of the dynamic properties of the involved protein–ligand complexes.

1.11. HARDLY AVOIDABLE: ENTHALPY/ENTROPY COMPENSATION

The phenomenon of enthalpy/entropy compensation has frequently been reported in protein–ligand complex series (Martin and Clements, 2013) and has been extensively studied in the binding of Lys-Xxx-Lys tripeptides to oligopeptide binding protein A (Sleigh et al., 1999; Davies et al., 1999). We studied a series of peptide mimetics with respect to trypsin binding (Brandt et al., 2011). All ligands exhibit a P_1 benzamidine anchor and a proline residue at P_2. The terminal part, intended to occupy the S_3 pocket, is composed of a cycloalkyl moiety of different ring size linked via an amino, ether, or methylene bridge to the proline portion (Fig. 1.9A). Hardly any differences in the Gibbs free energy are observed; however, partitioning in enthalpy and entropy shows large variations. Crystal structure analysis of the series with trypsin reveals remarkable deviations, not only in the binding modes but also in the crystal forms observed for the crystallized trypsin–ligand complexes. Two forms are observed, which differ in the mutual protein packing next to the S_3 pocket. The orthorhombic form shows rather large unoccupied space next to the pocket, whereas the packing in a trigonal form restricts accessible space quite strongly in the neighborhood of the S_3 pocket by adjacent protein molecules. Interestingly, ligands exhibiting the amino group linker prefer the open orthorhombic form and a disorder across at least two conformer families is found. Of these, one cluster resides with the cycloalkyl moiety in the S_3 pocket not forming a hydrogen bond to the adjacent Gly216CO. The alternative conformer family places the cycloalkyl ring outside the pocket and establishes the H-bond. The example highlights again the competition between favorable hydrophobic interactions and hydrogen bonding. Molecular dynamics (MD) simulations performed with these ligands confirm swapping between inside and outside orientation of the cycloalkyl portion to varying extent depending on the size and the conformational properties of the terminal ring. An increase of the entropic contributions for the ligands with larger terminal rings is experienced, which is supported by MD simulations. For the five-membered ring derivative, the binding mode outside the S_3 pocket seems more favorable and the hydrogen bond is formed. By contrast, the cyclohexyl moiety of the size-increased analog remains bound in the S_3 pocket during the simulation. It therefore lacks formation of the H-bond, yielding an entropically

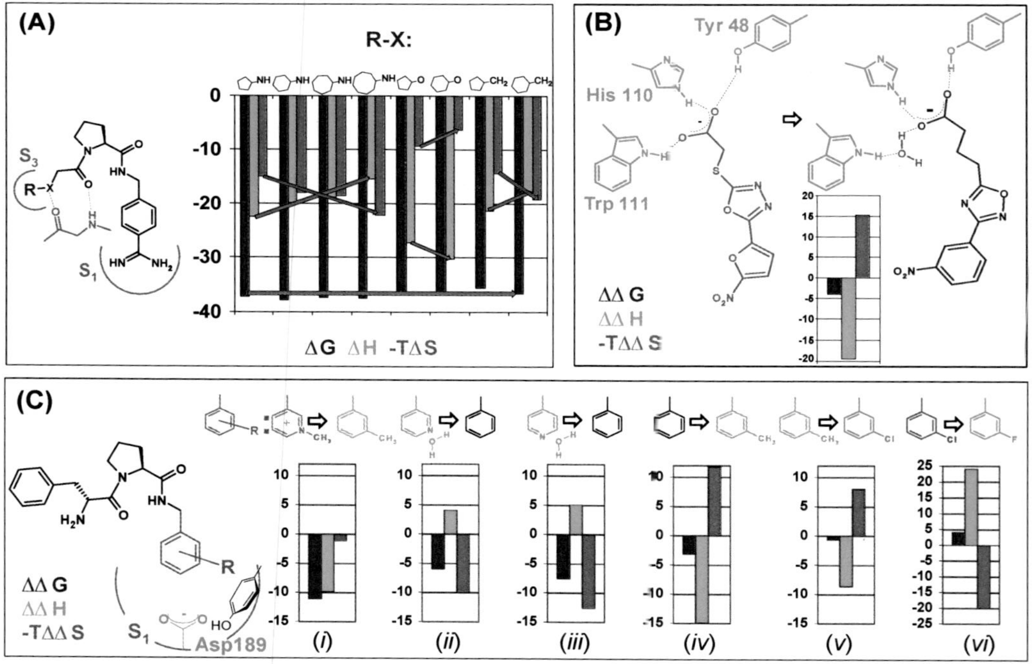

FIGURE 1.9 (A) A series of P_3-modified peptidomimetic ligands shows equipotent binding to trypsin nonetheless, affinity factorizes quite differently into enthalpy and entropy. Crystallography and MD simulations indicate differences in the residual mobility of the bound ligands, which explains, along with a competition between H-bonding and lipophilic contacts, the pronounced enthalpy/entropic compensation in this congeneric ligand series. (B) Two closely related aldose reductase inhibitors bind either with (right) or without (left) a water molecule mediating an H-bond between the ligand's charged carboxylate group and Trp111NH, His110, and Tyr48OH. A slight affinity improvement is observed for the ligand binding via the interstitial water molecule, resulting from a large exothermic and entropically disfavored signal. (C) Binding of a series of D-Phe-L-Pro-X ligands to thrombin has been analyzed by modifying the terminal P_1 aromatic moiety: (*i*) N-methyl pyridinium by toluolyl replacement wins significantly in ΔG due to the enthalpically less costly desolvation of the uncharged group. (*ii*, *iii*) Pyridine moieties in this position fix a water molecule next to Asp189, therefore their replacement by phenyl is entropically highly favorable. (*iv*) Attachment of a methyl group to the P_1 phenyl substituent improves affinity as an enthalpically favored contact to Tyr228 is formed and the size-increased toluolyl group reduces the residual mobility of the P_1 substituent in entropically unfavorable way. In addition, a fairly mobile water molecule is displaced, again a process beneficial in enthalpy but unfavorable in entropy. (*v*) Methyl-to-chlorine replacement at the P_1-phenyl ring gains little in ΔG, but strong enthalpic effects are experienced as chlorine interacts via its positively polarized tip with the negatively polarized aromatic moiety of Tyr225. (*vi*) Fluorine instead of chorine lacks this favorable electrostatic interaction to Tyr225 and due to the smaller size of fluorine, the P_1 substituent shows enhanced mobility, which is of entropic advantage.

favored profile owing to enhanced residual mobility. The derivatives with a seven- and eight-membered ring adopt more than one preferred in/out orientation, which are frequently interconverted. These enhanced intrinsic dynamics explain their increasing entropic signature.

By contrast, the ether- and methylene-linked derivatives prefer the closed trigonal form where ligand conformation is virtually restricted to one pose. Here, the hydrophobic pocket accommodates the terminal cycloalkyl moieties and no geometry is observed, which is able to establish a hydrogen bond to Gly216CO. Binding with less residual mobility most likely explains the more enthalpically favored binding profiles of the ether derivatives. Interestingly, the methylene- and amino-linked cyclopentyl derivatives exhibit similar thermodynamic profiles, but for different reasons. In the CH_2-linked ligand, the cyclopentyl portion resides permanently inside the S_3 pocket along the MD trajectory, whereas the amino analog remains outside. This suggests balanced contributions of hydrogen bonding and hydrophobic contacts in these derivatives.

This example shows the intimate correlation of hydrogen bonding and lipophilic contact formation, giving rise to rather different dynamic properties of the complexes. These determine the overall partitioning in enthalpic and entropic contributions, which virtually compensate in the example and make affinity enhancements next to the S_3 pocket very difficult.

1.12. WATER AND ITS IMPACT ON THE THERMODYNAMIC SIGNATURE

Capturing or releasing water molecules from binding sites may have a significant impact on the thermodynamic signature of ligand binding. Owing to enthalpy/entropy compensation, the effect on the free energy of binding is usually much smaller. The binding of sorbitol has been studied against the wild type and a Leu300 Pro variant of aldose reductase (Fig. 1.9B) (Petrova et al., 2005). Remarkably, the binding affinity to both protein species is nearly identical; however, enthalpy and entropy vary strongly. With its ether oxygen, sorbinil forms a water-mediated hydrogen bond to Leu300NH. As the NH group is missing in the mutant, also the interstitial water molecule is absent in the corresponding complex. Binding to the wild type is enthalpically more favorable but entropically more expensive than with the Pro mutant. An enthalpic advantage of -5 kJ/mol can be assigned to the water-mediated hydrogen bond that is formed. However, the uptake of the water molecule by the wild type is entropically costly. On the other hand, the Pro variant does not pay the entropic cost of entrapping a water molecule, but it also does not win the favorable enthalpy gain as no additional hydrogen bonds are formed.

A very similar enthalpy/entropy inventory is recorded for the displacement of ordered water molecules from the S_3 pocket of thrombin, again without substantial affinity enhancement (Biela et al., 2012a). As soon as charges are involved, water-mediated hydrogen bonds can achieve a contribution in the Gibbs free energy, even though part of the enthalpic benefit is diminished by the entropic costs for water capturing (Fig. 1.9B) (Steuber et al., 2007a).

We carefully analyzed the binding of P_1-modified tripeptide analogs to thrombin. In the remaining positions (P_2, P_3) all ligands exhibited an unchanged D-Phe-L-Pro scaffold. Replacement of an N-methylated pyridyl portion by its *m*-Me-phenyl analog results in a highly conserved binding mode (Biela et al., 2012c). Thus, the affinity differences can largely be ascribed to the desolvation costs of the strongly charged pyridinium moiety when compared with the geometrically equivalent toluoyl substituent. A huge loss in binding free energy is experienced, mainly paid by enthalpy (Fig. 1.9C*i*).

In the complexes of the *m*- and *p*-pyridyl derivatives water molecules are observed to mediate H-bonds between the pyridine nitrogen and the buried carboxylate group of Asp189, whereas in the structure with the unsubstituted phenyl analog these waters are displaced (Fig. 1.9C*ii,iii*). Nonetheless, the latter unsubstituted derivative shows higher potency predominantly for entropic reasons. This can be explained by the release of previously well-ordered water molecules to the bulk phase along with a significantly enhanced residual mobility of the P_1 ligand portion of the unsubstituted derivative in the S_1 pocket. In part, the observed entropic advantage is compensated by the endothermic desolvation costs of the more polar pyridyl derivatives.

The release of individual water molecules can also result in an overall more enthalpic signature (Baum et al., 2009b). Changing the P_1 substituent from phenyl to *m*-toluolyl entails the release of a water molecule from the S_1 pocket (Fig. 1.9C*iv*). Surprisingly, this displacement improves affinity due to an enthalpic gain, partly compensated by an entropic loss. Attachment of the methyl group reduces residual mobility in the binding pocket, which argues for an entropic loss. Nevertheless, the water release should be entropically favored as similarly observed in the previous case. The difference electron density assigned to this water molecule suggests either enhanced mobility or reduced occupancy in the S_1 pocket. Release of such a water molecule will gain entropically less as it is already quite mobile. By contrast, it can win enthalpically as better H-bonding contacts can be experienced in the bulk phase. Interestingly, this water molecule was difficult to predict and a reduced occupancy is suggested by a computational WaterMap analysis (Young et al., 2007; Abel et al., 2008). Thus this latter approach, which is based on an entirely different concept, comes to a very similar conclusion (Abel et al., 2011). The enthalpic signature of the toluoyl derivative is additionally enhanced by the attached methyl group, which undergoes a favorable van der Waals contact to the neighboring Tyr228. This effect can even be improved by replacement with an isosteric chlorine atom at this position (Fig. 1.9C*v*). Chlorine experiences attractive electrostatic interactions via its partial positive charge at the tip ("σ hole") with the negatively polarized center of the adjacent aromatic moiety and, thereby, reduces the residual mobility of the P_1 substituent even further. Remarkably, the *m*-F to *m*-Cl replacement is characterized by an affinity loss due to an enthalpic disadvantage and an entropic gain (Fig. 1.9C*vi*). Fluorine does not exhibit the positive tip for attractive electrostatic interactions. Due to its smaller size, augmented residual mobility is observed for the fluorine derivative leading to an entropic benefit.

1.13. IMPACT OF SURFACE WATER MOLECULES ON THE THERMODYNAMIC SIGNATURE OF PROTEIN–LIGAND COMPLEXES

Proteins are wetted in aqueous solution and generate layers of water molecules around their surface. If a new protein–ligand complex is formed and the ligand contributes with some parts of its substituents, particularly in flat, surface-exposed binding pockets to the newly formed common surface, the generated water surface layer will change. This change has a surprisingly large impact on the thermodynamic signature of the formed complex. We studied this phenomenon extensively in a series of congeneric P_2'-varied thermolysin complexes (Biela et al., 2012b, 2013; Krimmer et al., 2014). In all complexes, the binding mode of the ligands remains virtually unchanged apart from the arrangement of the water molecules adjacent to the P_2' substituents. At first, a congeneric set of four inhibitors should be considered (Fig. 1.10). Attachment of a methyl group to the parent scaffold increases the affinity by about −2 kJ/mol, a value expected for a favorably placed methyl group, which factorizes slightly into a beneficial change in enthalpy and unfavorable change in entropy. As the methyl group requires additional space, two water molecules present in the complex with the nonmethylated derivative are displaced but two new ones are captured by the newly formed complex. Adding a carboxylate group to this scaffold results in a pronounced affinity gain combined with large mutually compensating exothermic and entropically unfavorable effects. Such a profile is expected for the attachment of a polar group, well suited to establish strong enthalpic interactions with its environment but reducing simultaneously the residual mobility of the arrangement in the local neighborhood. If the attachments are performed with inverted sequence, the observed effects reverse their order with different magnitude. Now the carboxylate attachment yields only about −1 kJ/mol in Gibbs free energy, with minor partitioning in enthalpy and entropy. Therefore, sole explanations in terms of desolvation costs of the strongly perturbing carboxylate group are not appropriate. It is more surprising that the subsequently added methyl group induces a strong affinity increase concatenated with an extensive enthalpy/entropy compensation. Three of the complexes show a contiguously connected water network that wraps around the terminal hydrophobic substituent and involves, if present, the carboxylate group. In the complex with the ligand solely exhibiting the carboxylate but lacking the methyl group, the water

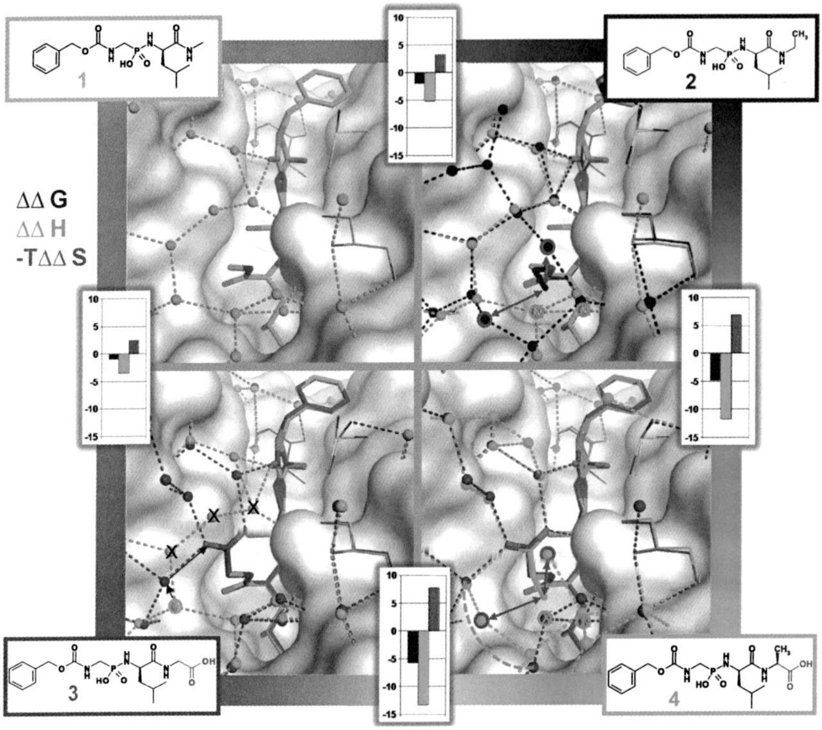

FIGURE 1.10 Ligand **1** (orange, upper left) lacking methyl and carboxylate group binds to thermolysin (TLN) establishing an extensive water network wrapping around the terminal P_2' group. Upon methylation **2** is produced (blue, upper right, superimposed with TLN-**1**) and a small enthalpically favored affinity gain is experienced. Two water molecules (encircled and crossed in blue) are displaced, whereas two new ones are entrapped (red), which form favorable contacts to the newly attached methyl group. Modifying **2** further to **4** (green, lower right, superimposed with TLN-**3**) by adding the charged carboxylate group results in a strong affinity gain overwhelmingly for enthalpic reasons. Surprisingly, attachment of the carboxylate to **1** forming **3** (magenta, lower left, superimposed with TLN-**1**) is responded by a very weak thermodynamic signature even though three water molecules (black crosses) are replaced by the carboxylate group, and one water molecule (cyan) is shifted to enable a contact with the introduced COO^- group. Finally adding the methyl group to **3** featuring **4** recovers the thermodynamic profile expected for the carboxylate attachment. Remarkably, in TLN-**3** the water network wrapping around the P_2' substituent is broken (indicated by the dashed line), which is responded by a loss in affinity for enthalpic reasons. Instead, entropy is less affected as the broken network sacrifices less residual mobility. In TLN-**4** (green) the water network is sealed again and the sole methyl attachment rescues the entire affinity gain with enthalpically favored signature.

network is broken and appears incomplete. This rupture should be enthalpically unfavorable and entropically beneficial, as it captures fewer water molecules and the broken network allows for larger residual mobility. This counterbalances favorable contributions resulting from the introduction of the polar carboxylate group. If the additional methyl group is subsequently introduced, this contribution is rescued as the water network is reestablished and contiguously wraps around the methyl group. Therefore, an additional water molecule is recruited, which fills the gap in the network. Its binding is favorably stabilized by van der Waals interactions to the added methyl group of the ligand.

Attachment of the terminal carboxylate group recruits a water molecule at the capping position on top of the charged acidic group. Occupancy of this position provides an enthalpic advantage to the formed complex, as evident from the thermodynamic profiles of the ligands exhibiting the terminal carboxylate group and P_2' substituents of varying size (Fig. 1.11, left) (Biela et al., 2013). Four of the complexes show a significant enthalpic advantage over the others. They all accommodate the capping water molecule. Only the unsubstituted Gly analog cannot stabilize this water binding site and the derivatives with larger P_2' side chains displace this water molecule as a consequence of steric repulsion.

In the current series, the attachment of an increasingly hydrophobic substituent results in quite a difference in the enthalpic and entropic responses.

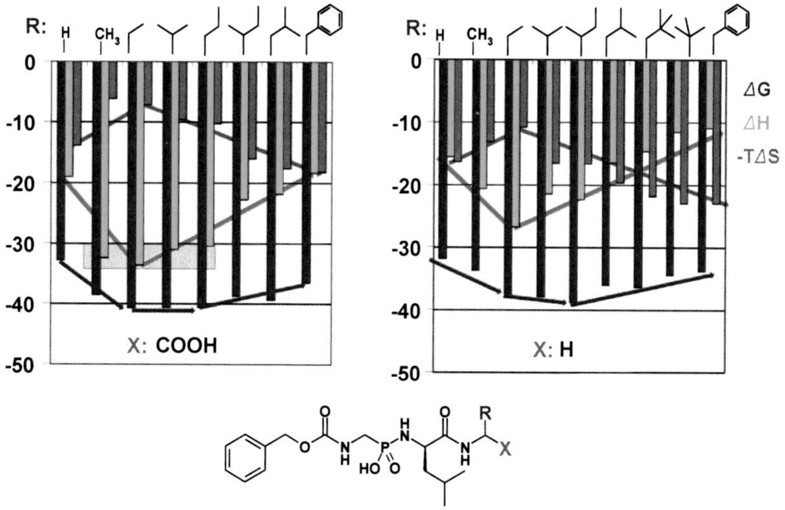

FIGURE 1.11 Two congeneric series of thermolysin inhibitors with P_2' substituents of growing hydrophobicity (R) with (left) and without (right) terminal carboxylate group (X) all exhibit a conserved binding mode. Differences result from modulations of the local solvation structure next to the S_2' pocket. Both series show similar trends in the thermodynamic signature with best affinity for small- to medium-sized substituents. For these enthalpy dominates ΔG; with larger substituents the entropic component overwhelms. Even though the thermodynamic profiles follow similar patterns, the structural details determining the signatures are quite different particularly with respect to the quality and perfection the water network formed next to the P_2' substituent. For example, for the X=COOH series four complexes show favored enthalpic signature (*gray box*). They all exhibit a water molecule in capping position on top of the charged carboxylate groups contributing favorable electrostatic interactions. The other complexes lack this extra water molecule.

Usually the placement of hydrophobic substituents into hydrophobic protein pockets is described as the "hydrophobic effect" and is quantified by hydrophobic surface patches, which are removed from solvent access. The current example shows that the hydrophobic effect has to be described as a structural phenomenon and relates to changes in the local water structure. Thus, models simply focusing on surface patches removed from solvent exposure are not sufficient and cannot explain the trends in the series. This phenomenon becomes even more obvious in the ligand series modulating the hydrophobic P_2' substituent but lacking the terminal carboxylate group (Krimmer et al., 2014). Here small- to medium-sized substituents enhance binding affinity predominantly for enthalpic reasons (Fig. 1.11, right). This relates to an improvement and stepwise completion of the water network established at the surface of the newly formed complexes. For larger substituents more complete filling of the S_2' pocket is achieved, but they prevent a complete and unperturbed water network to be built. An increasing number of water molecules is displaced, which is reflected by a growing entropic signal and a reduced binding affinity.

Interestingly, a single methyl group can shift the thermodynamic profile from more enthalpy to more entropy driven. Again, the difference relates to a change in the local water structure. The enthalpically favored complex, which features one methyl group less, shows a rather complete water network that wraps around the newly created protein–ligand complex surface. Surprisingly, the sole addition of an extra methyl group leads to an extensive removal of surface water molecules. Owing to the increased volume requirements of the added methyl group several water molecules, assembled across the surface of the complex with the smaller P_2' substituent, are repelled from their previous positions and the newly formed complex loses part of its favorable surface solvation layer. This water displacement leads to an entropic signal dominating the signature for the formation of this complex.

1.14. CONCLUSION

Drug development based on rational concepts requires detailed understanding of the interactions of a small molecule drug with its target protein. Correlation of structural and thermodynamic properties of ligand–protein binding in terms of enthalpy/entropy profiles provide conclusive insights. It has been proposed to use such profiles to support the decision-making process regarding which ligands to take as lead candidates to the next level of development (Freire, 2008; Ladbury et al., 2010; Olsson et al., 2011; Chodera and Mobley, 2013; Dunitz, 2003; Murray and Verdonk, 2002; Nazare et al., 2012; Borsi et al., 2010; Olsson et al., 2008; Freire, 2009; Hann and Keserü, 2011; Ferenczy and Keserü, 2010, 2012, 2016; Reynolds and Holloway, 2011; Martin and Clements, 2013). From a theoretical point of view, it appears promising and advisable to focus on the most enthalpic binders, as optimization steps governed by entropic factors will be followed unavoidably during late stage optimization. Prerequisite to apply such strategies is the correct reasoning why a given protein–ligand complex exhibits a particular thermodynamic binding signature, as only then the selection of the lead with the "largest enthalpic efficiency" will be useful (Ladbury et al., 2010).

The method of choice to record thermodynamic data is ITC. It provides direct access to ΔG and ΔH in one single experiment at constant temperature; TΔS, however, is calculated from their numerical difference. This causes an inevitable ΔH/TΔS compensation, apart from an intrinsic enthalpy/entropy compensation given in biological systems. To avoid being trapped in error-prone compensatory measurements, thorough analysis and correction of superimposed systematic errors affecting ITC data have to be performed. Since differences in the applied salt concentrations, added detergents, cofactors, or co-substrates, all influence the enthalpy/entropy partitioning, it is highly advisable to *only* correlate the series of matching protein–ligand pairs relative to each other, when measured under similar conditions. In such case, the relative differences are hardly affected. Even though frequently attempted, global comparisons across large data samples extracted from literature must fail due to problems that such heterogeneous data are hardly scaled to a common reference state. Unfortunately, the heat signal measured in an ITC experiment encompasses all changes involving the entire protein–ligand binding event. This makes factorization and subsequent assignment of the profile changes to individual interactions or functional group contributions extremely difficult and additional information, particularly about the structural properties of the formed complexes, is highly desirable.

Nevertheless, data correlated across congeneric series provide important insights. Hydrogen bonding usually results in an enthalpic signal, which enhances once growing charges of the interacting functional groups are involved. However, with increasing charges a detrimental entropic contribution is also experienced, which reduces, due to enthalpy/entropy compensation, the overall free energy contribution of an H-bond. Lipophilic contacts buried upon complex formation result in an increasing entropic signal, but only if ordered water molecules are displaced from the binding pocket. Displacement of mobile water molecules upon ligand binding can also give rise to a more enthalpy-driven binding (Homans, 2007). Furthermore, poorly or insufficiently solvated binding pockets can result in high-affinity binding of entirely hydrophobic ligand portions resulting in a more enthalphically favored signature (Englert et al., 2010). If no permanent and strong charges of the interacting species are involved, the release or entrapment of water molecules upon ligand binding seems to be virtually balanced out in the Gibbs free energy inventory, but huge effects are experienced with respect to the enthalpy/entropy partitioning. This observation demonstrates that the sole determination of the free energy (what is usually done in an assay characterizing "affinity") will hardly unravel involvement of water molecules in binding. This also explains why surprisingly many computer-modeling approaches can generate reasonable ΔG predictions, even though they neglect water molecules. As a consequence, the predicted binding poses will be incorrect, a fact usually only recognized if crystal structure analysis is performed in parallel or a given binding mode is used as the basis for further ligand design.

Ligand preorganization and rigidization of the protein-bound conformation may result in large beneficial free energy contributions, mainly due to an entropic advantage. These generalized signatures often become only transparent once a congeneric series of ligands is evaluated as the overall thermodynamic profile of the binding process results as a sum of multiple and superimposed effects. They originate from changes in the dynamics of protein and/or ligand, rearrangements of the protein, and most important by changes of the solvation pattern of discrete water molecules. Furthermore, puzzling cooperativity between hydrogen bonding and hydrophobic contacts could lead to changes in the dynamics of protein–ligand complexes (MacRaild et al., 2007; Diehl et al., 2009; Popovych et al., 2006; Zidek et al., 1999; Diehl et al., 2010; Stöckmann et al., 2008; Syme et al., 2010; Falconer, 2016) or modulations of residual solvation patterns (Biela et al., 2012b, 2013).

Every formed protein–ligand complex creates a new common surface, which is wetted by the surrounding solvent. The quality and perfection of the formed surface water network that wraps especially around partly exposed substituents of the bound ligand strongly influences the affinity and thermodynamic signature of the formed complex. Ideal fit results in affinity enhancement of the bound ligand; imperfect and fragmented water networks reduce affinity of the bound ligand. As these modulations of the surface water network result directly from the properties of the bound ligand and its partly solvent-exposed substituents, well-established medicinal chemistry optimization steps can accomplish a fine-tuning of the thermodynamic properties. Moreover, such changes are reflected by major modulations of the enthalpy/entropy signature and easily provoke a mutual ΔH versus $T\Delta S$ shift of several kilojoules per moles. If the residual solvation pattern takes such an enormous impact on the thermodynamic signature of a formed complex, any classification of a given ligand as "more enthalpic" or "more entropic" binder appears rather useless without having full structural information about the formed complex, e.g., by means of high-resolution crystal structure analyses. Only then, the thermodynamic profile can support the decision regarding which ligand to take as candidate for optimization on the next level of development. Nonetheless, deviating thermodynamic profiles recorded across congeneric ligand series unambiguously indicate differences in the binding patterns, be it for deviations in the binding poses, the residual solvation patterns, or the intrinsic dynamics of the formed complexes.

Even though we can establish some general rules on how to combat enthalpy/entropy compensation and substantiate some reasoning why it might be beneficial to start with leads of "high enthalpic efficiency", the binding event as a whole shows many additional phenomena, giving rise to a mutual undesired enthalpy/entropy compensation. It remains questionable as to whether these effects can always be fully elucidated and subsequently avoided in an optimization strategy.

REFERENCES

Abel, R., Young, T., Farid, R., Berne, B.J., Friesner, R.A., 2008. Role of the active-site solvent in the thermodynamics of factor Xa ligand binding. J. Am. Chem. Soc. 130, 2817–2831.

Abel, R., Salam, N.K., Shelley, J., Farid, R., Freisner, R.A., Sherman, W., 2011. Contribution of explicit solvent effects to the binding affinity of small-molecule inhibitors in blood coagulation factor serine proteases. ChemMedChem 6, 1049–1066.

Ajay, Murcko, M.A., 1995. Computational methods to predict binding free energy in ligand-receptor complexes. J. Med. Chem. 38, 4953–4967.

Ali, A., Bandaranayke, R.M., Cai, Y., King, N.M., Kolli, M., Mittal, S., Murzycki, J.F., Nalam, M.N.L., Nalivaika, E.A., Özen, A., Prabu-Jeyabalan, M.M., Thayer, K., Schiffer, C.A., 2010. Viruses 2, 2509–2535.

Baker, B.M., Murphy, K.P., 1996. Evaluation of linked protonation effects in protein binding reactions using isothermal titration calorimetry. Biophys. J. 71, 2049–2055.

Baranauskienėm, L., Petrikaitė, V., Matulienė, J., Matulis, D., 2009. Crown ether study: titration calorimetry standards and the precision of isothermal titration calorimetry data. Int. J. Mol. Sci. 10, 2752–2762.

Baum, B., Muley, L., Heine, A., Smolinski, M., Hangauer, D., Klebe, G., 2009a. Think twice: understanding the high potency of bis(phenyl)methane inhibitors of thrombin. J. Mol. Biol. 391, 552–564.

Baum, B., Mohamed, M., Zayed, M., Gerlach, C., Heine, A., Hangauer, D., Klebe, G., 2009b. More than a simple lipophilic contact: a detailed thermodynamic analysis of non-basic residues in the S_1 pocket of thrombin. J. Mol. Biol. 390, 56–69.

Baum, B., Muley, L., Smolinski, M., Heine, A., Hangauer, D., Klebe, G., 2010. Non-additivity of functional group contributions in protein-ligand binding: a comprehensive study by crystallography and isothermal titration calorimetry. J. Mol. Biol. 397, 1042–1057.

Biela, A., Sielaff, F., Terwesten, F., Heine, A., Steinmetzer, T., Klebe, G., 2012a. Ligand binding stepwise disrupts water network in thrombin: enthalpic and entropic changes reveal classical hydrophobic effect. J. Med. Chem. 55, 6094–6110.

Biela, A., Betz, M., Heine, A., Klebe, G., 2012b. Water makes the difference: rearrangement of water solvation layer triggers non-additivity of functional group contributions in protein-ligand binding. ChemMedChem 7, 1423–1434.

Biela, A., Khayat, M., Tan, H., Kong, J., Heine, A., Hangauer, A.D., Klebe, G., 2012c. Impact of ligand and protein desolvation on ligand binding to the S_1 pocket of thrombin. J. Mol. Biol. 418, 350–366.

Biela, A., Nasief, N.N., Betz, M., Heine, A., Hangauer, D., Klebe, G., 2013. Dissecting the hydrophobic effect on the molecular level: the role of water, enthalpy, and entropy in ligand binding to thermolysin. Angew. Chem. Int. Ed. 52, 1822–1828.

Blum, A., 2007 (PhD thesis). Univ. of Marburg, Germany.

Blundell, T.L., Jhoti, H., Abell, C., 2002. High-throughput crystallography for lead discovery in drug design. Nat. Rev. Drug Discov. 2, 45–53.

Borsi, V., Calderone, V., Fragai, M., Luchinat, C., Sarti, N., 2010. Entropic contribution to the linking coefficient in fragment-based drug design: a case study. J. Med. Chem. 53, 4285–4289.

Brandt, T., Holzmann, N., Muley, L., Khayat, M., Wegscheid-Gerlach, C., Baum, B., Heine, A., Hangauer, D., Klebe, G., 2011. Congeneric but still distinct: how closely related trypsin ligands exhibit different thermodynamic and structural properties. J. Mol. Biol. 405, 1170–1187.

Chaires, J.B., 2008. Calorimetry and thermodynamics in drug design. Ann. Rev. Biophys. 37, 135–151.

Chandler, D., 2005. Interfaces and the driving force of hydrophobic assembly. Nature 437, 640–647.

Cheng, Y.-C., Prusoff, W.H., 1973. Relationship between the inhibition constant (K_i) and the concentration of inhibitor which causes 50 per cent inhibition (IC_{50}) of an enzymatic reaction. Biochem. Pharmacol. 22, 3099–3108.

Chodera, J.D., Mobley, D.L., 2013. Entropy-Enthalpy compensation: role and ramifications in biomolecular ligand recognition and design. Ann. Rev. Biophys. 42, 121–142.

Czodrowski, P., Sotriffer, C.A., Klebe, G., 2007. Protonation changes upon ligand binding to trypsin and thrombin: structural interpretation based on pK_a calculations and ITC experiments. J. Mol. Biol. 367, 1347–1356.

Das, K., Lewi, P.J., Hughes, S.H., Arnold, E., 2005. Crystallography and the design of anti-AIDS drugs: conformational flexibility and positional adaptability are important in the design of non-nucleoside HIV-1 reverse transcriptase inhibitors. Prog. Biophys. Mol. Biol. 88, 209–231.

Davies, T.G., Hubbard, R.E., Tame, J.R.H., 1999. Relating structures to thermodynamics: the crystal structures and binding affinity of eight OppA-peptide complexes. Protein Sci. 8, 1432–1444.

Diehl, C., Genheden, S., Modig, K., Ryde, U., Akke, M., 2009. Conformational entropy changes upon lactose binding to the carbohydrate recognition domain of galectin-3. J. Biomol. NMR 45, 157–169.

Diehl, C., Engström, O., Delaine, T., Håkansson, M., Genheden, S., Modig, K., Leffler, H., Ryde, U., Nilsson, U.J., Akke, M., 2010. Protein flexibility and conformational entropy in ligand design targeting the carbohydrate recognition domain of galectin-3. J. Am. Chem. Soc. 132, 14577–14589.

Dill, K.A., Truskett, T.M., Vlachy, V., Hribar-Lee, B., 2005. Modeling water, the hydrophobic effect, and ion solvation. Annu. Rev. Biophys. Biomol. Struct. 34, 173–199.

Dunitz, J.D., 2003. Win some, lose some: enthalpy-entropy compensation in weak intermolecular interactions. Chem. Biol. 2, 709–712.

Englert, L., Biela, A., Zayed, M., Heine, A., Hangauer, D., Klebe, G., 2010. Displacement of disordered water molecules from the hydrophobic pocket creates enthalpic signature: binding of phosphonamidate to the S_1'-pocket of thermolysin. Biochim. Biophys. Acta 1800, 1192–1202.

Erlanson, D.A., McDowell, R.S., O'Brien, T., 2004. Fragment-based drug discovery. J. Med. Chem. 47, 3463–3482.

Falconer, R.J., 2016. Applications of isothermal titration calorimetry – the research and technical developments from 2011 to 2015. J. Mol. Recognit. 29, 504–515. http://dx.doi.org/10.1002/jmr.2550.

Falconer, R.J., Collins, B.M., 2011. Survey of the year 2009: applications of isothermal titration calorimetry. J. Mol. Recognit. 24, 1–16.

Fenwicka, R.B., Bedemb, H.v. d., Fraserc, J.S., Wright, P.E., 2014. Integrated description of protein dynamics from room-temperature X-ray crystallography and NMR. Proc. Natl. Acad. Sci. U.S.A. 111, E445–E454.

Ferenczy, G.G., Keserü, G.M., 2010. Thermodynamics guided lead discovery and optimization. Drug Discov. Today 15, 919–932.

Ferenczy, G.G., Keserü, G.M., 2012. Thermodynamics of fragment binding. J. Chem. Inf. Model. 52, 1039–1045.

Ferenczy, G.G., Keserü, G.M., 2013. How are fragments optimized? A retrospective analysis of 145 fragment optimizations. J. Med. Chem. 56, 2478–2486.

Ferenczy, G.G., Keserü, G.M., 2016. On the enthalpic preference of fragment binding. Med. Chem. Commun. 7, 332–337.

Fernandez, A., Frazer, C., Scott, L.R., 2012. Purposely engineered drug-target mismatches for entropy-based drug optimization. Trends Biotechn. 30, 1–7.

Fraser, J.S., van den Bendem, H., Samelson, A.J., Lang, P.J., Holton, J.M., Echols, N., Alber, T., 2011. Accessing protein conformational ensembles using room-temperature X-ray crystallography. Proc. Nat. Acad. Sci. U.S.A. 108, 16247–16252.

Freire, E., 2008. Do enthalpy and entropy distinguish first in class from best in class? Drug Discov. Today 13, 869–874.

Freire, E., 2009. A thermodynamic approach to the affinity optimization of drug candidates. Chem. Biol. Drug Des. 74, 472–488.

Glas, A., Bier, D., Hahne, G., Rademacher, C., Ottmann, C., Grossmann, T.N., 2014. Constrained peptides with target-adapted cross-inks as inhibitors of a pathogenic protein-protein interaction. Angew. Chem. Int. Ed. 53, 2489–2493.

Goldberg, R.N., Kishore, N., Lennen, R.M., 2002. Thermodynamic quantities for the reactions of buffers. J. Phys. Chem. Ref. Data 31, 231–370.

Hann, M.M., Keserü, G.M., 2011. Finding the sweet spot: the role of nature and nurture in medicinal chemistry. Nat. Rev. Drug Discov. 11, 355–365.

Homans, S.W., 2005. Probing the binding entropy of ligand protein interactions by NMR. ChemBioChem 6, 1–8.

Homans, S.W., 2007. Water, water everywhere – except where it matters. Drug Discov. Today 12, 534–539.

Hopkins, A.L., Groom, C.R., Alex, A., 2004. Ligand efficiency: a useful metric for lead selection. Drug Discov. Today 9, 430–431.

Horn, J.R., Russell, D., Lewis, E.A., Murphy, K.P., 2001. Van't Hoff and calorimetric enthalpies from ITC: are there significant discrepancies? Biochemistry 40, 1774–1778.

Huschmann, F.U., Linnik, J., Sparta, K., Ühlein, M., Wang, X., Metz, A., Schiebel, J., Heine, A., Klebe, G., Weiss, M.S., Mueller, U., 2016. Structures of endothiapepsin–fragment complexes from crystallographic fragment screening using a novel, diverse and affordable 96-compound fragment library. Acta Cryst. F72, 346–355.

Jelesarov, I., Bossard, H.R., 1999. Isothermal titration calorimetry and differential scanning calorimetry as complementary tools to investigate the energetic of biomolecular recognition. J. Mol. Recognit. 12, 3–18.

Kasinath, V., Sharp, K.A., Wand, A.J., 2013. Microscopic insights into the NMR relaxation-based protein conformational entropy meter. J. Am. Chem. Soc. 135, 15092–15100.

Keserü, G.M., Erlanson, D.A., Ferenczy, G.G., Hann, M.M., Murray, C.W., Pickett, S.D., 2016. Design principles for fragment libraries: maximizing the value of learnings from pharma fragment-based drug discovery (FBDD) programs for use in academia. J. Med. Chem. 59, 8189–8206. http://dx.doi.org/10.1021/acs.jmedchem.6b00197.

Klebe, G., 2013. Drug Design. Springer Reference, Heidelberg, New York, Dordrecht, London (Chapter 4).

de Kloe, G.E., Bailey, D., Leurs, R., de Esch, I.J.P., 2009. Transforming fragments into candidates: small becomes big in medicinal chemistry. Drug Discov. Today 14, 630–646.

Krimmer, S., Klebe, G., 2015. Thermodynamics of protein–ligand interactions as a reference for computational analysis: how to assess accuracy, reliability and relevance of experimental data. J. Comput. Aided Molec. Des. 29, 867–883.

Krimmer, S.G., Betz, M., Heine, A., Klebe, G., 2014. Methyl, ethyl, propyl, butyl: futile but not for water, as the correlation of structure and thermodynamic signature shows in a congeneric series of thermolysin inhibitors. ChemMedChem 9, 833–846.

Kyte, J., 2003. The basis of the hydrophobic effect. Biophys. Chem. 100, 193–203.

Ladbury, J.E., 2001. Isothermal titration calorimetry: application to structure-based drug design. Thermochim. Acta 380, 209–215.

Ladbury, J.E., Chowdhry, B.Z., 1996. Sensing the heat: the application of isothermal titration calorimetry to thermodynamic studies of biomolecular interactions. Chem. Biol. 3, 791–801.

Ladbury, J.E., Klebe, G., Freire, E., 2010. Adding calorimetric data to decision making in lead discovery: a hot tip. Nat. Rev. Drug Discov. 9, 23–27.

Liu, Y., Sturtevant, J.M., 1995. Significant discrepancies between van't Hoff and calorimetric enthalpies II. Protein Sci. 4, 2559–2561.

MacRaild, C.A., Daranas, A.H., Bronowska, A., Homans, S.W., 2007. Global changes in local protein dynamics reduce the entropic cost of carbohydrate binding in the arabinose binding protein. J. Mol. Biol. 368, 822–832.

Martin, S.F., Clements, J.H., 2013. Correlating structure and energetics in protein-ligand interactions: paradigms and paradoxes. Annu. Rev. Biochem. 82, 267–293.

Mizoue, L.S., Tellinghuisen, J., 2004. Calorimetric vs. van't Hoff binding enthalpies form ITC: Ba^{2+} - crown ether complexation. Biophys. Chem. 110, 15–24.

Mondal, M., Radeva, N., Köster, H., Park, A., Potamitis, C., Zervou, M., Klebe, G., Hirsch, A.K.H., 2014. Structure-based design exploiting dynamic combinatorial chemistry to identify novel inhibitors for the aspartic protease endothiapepsin. Angew. Chem. Int. Ed. 53, 3259–3263.

Muley, L., Baum, B., Smolinski, M., Freindorf, M., Heine, A., Klebe, G., Hangauer, D., 2010. Enhancement of hydrophobic interactions and hydrogen bond strength by cooperativity: synthesis, modeling, and molecular dynamics simulations of a series of thrombin inhibitors. J. Med. Chem. 53, 2126–2135.

Murray, C.W., Verdonk, M.L., 2002. The consequences of translational and rotational entropy lost by small molecules on binding to proteins. J. Comput. Aided Molec. Des. 16, 741–753.

Myszka, D.G., Abdiche, Y.N., Arisaka, F., Byron, O., Eisenstein, E., Hensley, P., Thomson, J.A., Lombardo, C.R., Schwarz, F., Stafford, W., Doyle, M.L., 2003. The ABRF-MIRG'02 study: assembly state, thermodynamic, and kinetic analysis of an enzyme/inhibitor interaction. J. Biomol. Techn. 4, 247–269.

Nakasako, M., 2004. Water-protein interactions from high-resolution protein crystallography. Philos. Trans. R. Soc. London Ser. B 359, 1191–1206.

Nazare, M., Matter, H., Will, D.W., Wagner, M., Urmann, M., Czech, J., Schreuder, H., Bauer, A., Ritter, K., Wehner, V., 2012. Fragment deconstruction of small, potent factor Xa inhibitors: exploring the superadditivity energetic of fragment linking in protein-ligand complexes. Angew. Chem. Int. Ed. 51, 905–911.

Neeb, M., Czodrowski, P., Heine, A., Barandun, L.J., Hohn, C., Diederich, F., Klebe, G., 2014. Chasing protons: how ITC, mutagenesis and pK_a calculations trace the locus of charge in ligand binding to a tRNA-binding enzyme. J. Med. Chem. 57, 5554–5565.

Neeb, M., Hohn, C., Ehrmann, F.R., Härtsch, A., Heine, A., Diederich, F., Klebe, G., 2016. Occupying a flat subpocket in a tRNA-modifying enzyme with ordered or disordered side chains: favorable or unfavorable for binding? Bioorg. Med. Chem. 24, 4900–4910. BMC-D-16–00702 accepted.

Ohtaka, H., Freire, E., 2005. Adaptive inhibitors of the HIV-1 protease. Prog. Biophys. Mol. Biol. 88, 193–208.

Olsson, T.S.G., Williams, M.A., Pitt, W.R., Ladbury, J.E., 2008. The thermodynamics of protein-ligand interactions and solvation: insights for ligand design. J. Mol. Biol. 384, 1002–1017.

Olsson, T.S.G., Ladbury, J.E., Pitt, W.R., Williams, M.A., 2011. Extent of enthalpy-entropy compensation in protein-ligand interactions. Protein Sci. 20, 1607–1618.

Pethica, B.A., 2015. Misuse of thermodynamics in the interpretation of isothermal titration calorimetry data for ligand binding to proteins. Anal. Biochem. 472, 21–29.

Petrova, T., Steuber, H., Hazemann, I., Cousido-Siah, A., Mitschler, A., Chung, R., Oka, M., Klebe, G., El-Kabbani, O., Joachimiak, A., Podjarny, A., 2005. Factorizing selectivity determinants of inhibitor binding towards aldose and aldehyde reductases: structural and thermodynamic properties of the aldose reductase mutant Leu300Pro-fidarestat complex. J. Med. Chem. 4, 5659–5665.

Popovych, N., Sun, S., Ebright, R.H., Kalodimos, C.G., 2006. Dynamically driven protein allostery. Nat. Struct. Biol. 13, 831–838.

Radeva, N., Schiebel, J., Wang, X., Krimmer, S.G., Fu, K., Stieler, M., Ehrmann, F.R., Metz, A., Rickmeyer, T., Betz, M., Winquist, J., Park, A.Y., Huschmann, F.U., Weiss, M., Mueller, U., Heine, A., Klebe, G., 2016a. Active site mapping of an aspartic protease by multiple fragment crystal structures: versatile warheads to address a catalytic dyad. J. Med. Chem. http://dx.doi.org/10.1021/acs.jmedchem.6b01195.

Radeva, N., Krimmer, S.G., Stieler, M., Schiebel, J., Fu, K., Wang, X., Ehrmann, F.R., Metz, A., Huschmann, F.U., Weiss, M., Mueller, U., Heine, A., Klebe, G., 2016b. Remote interplay of small molecules with endothiapepsin – hot spot analysis. J. Med. Chem. 59, 7561–7575.

Rees, D.C., Congreve, M., Murray, C.W., Carr, R., 2004. Fragment-based lead discovery. Nat. Rev. Drug Discov. 3, 660–672.

Reynolds, C.H., Holloway, M.K., 2011. Thermodynamics of ligand binding and efficiency. ACS Med. Chem. Lett. 2, 433–437.

Rühmann, E., Betz, M., Fricke, M., Heine, A., Schäfer, M., Klebe, G., 2015a. Thermodynamic signatures of fragment binding: validation of direct versus displacement ITC titrations. Biochim. Biophys. Acta 1850, 647–656.

Rühmann, E., Betz, M., Heine, A., Klebe, G., 2015b. Fragment binding can be either more enthalpy-driven or entropy-driven: crystal structures and residual hydration patterns suggest why. J. Med. Chem. 58, 6960–6967.

Rühmann, E., Rupp, M., Betz, M., Heine, A., Klebe, G., 2016. Boosting affinity by correct ligand preorganization for the S_2 pocket of thrombin: a study by ITC, MD and high resolution crystal structures. ChemMedChem 11, 309–319.

Schiebel, J., Radeva, N., Krimmer, S.G., Wang, X., Stieler, M., Ehrmann, F.R., Fu, K., Huschmann, F.U., Metz, A., Huschmann, F.U., Weiss, M.S., Mueller, U., Heine, A., Klebe, G., 2016. Six biophysical screening methods miss a large proportion of crystallographically discovered fragment hits: a case study. ACS Chem. Biol. 11, 1693–1701.

Setny, P., Baron, R., McCammon, J.A., 2010. How can hydrophobic association be enthalpy driven? J. Chem. Theory Comput. 6, 2866–2871.

Sharp, K., 2001. Entropy-enthalpy compensation: fact or artifact? Protein Sci. 10, 661–667.

Simunec, J., 2007 (Ph.D. thesis). Univ. of Marburg.

Sleigh, S.H., Seavers, P.R., Wilkingson, A.J., Ladbury, J.E., Tame, J.R.H., 1999. Crystallographic and calorimetric analysis of peptide binding to OppA protein. J. Mol. Biol. 291, 393–415.

Snyder, P.W., Mecinovic, J., Moustakas, D.T., Thomas III, S.W., Harder, M., Lockett, M.R., Heroux, A., Sherman, W., Whitesides, G.M., 2011. Mechanism of the hydrophobic effect in the biomolecular recognition of arylsulfonamides by carbonic anhydrase. Proc. Natl. Acad. Sci. U.S.A. 108, 17889–17894.

Stegmann, C.M., Seeliger, D., Sheldrick, G.M., de Groot, B., Wahl, M.C., 2009. Thermodynamic signature of trapped water molecules in a protein-ligand interaction. Angew. Chem. Int. Ed. 48, 5207–5210.

Steuber, H., Heine, A., Klebe, G., 2007a. Structural and thermodynamic study on aldose reductase: nitro-substituted inhibitors with strong enthalpic binding contribution. J. Mol. Biol. 368, 618–638.

Steuber, H., Czodrowski, P., Sotriffer, C.A., Klebe, G., 2007b. Tracing changes in protonation: a prerequisite to factorize thermodynamic data of inhibitor binding to aldose reductase. J. Mol. Biol. 373, 1305–1320.

Stöckmann, H., Bronowska, A., Syme, N.R., Thompson, G.S., Kalverda, A.P., Stuart, L., Warriner, S.L., Homans, S.W., 2008. Residual ligand entropy in the binding of *p*-substituted benzenesulfonamide ligands to bovine carbonic anhydrase II. J. Am. Chem. Soc. 130, 12420–12426.

Syme, N.R., Dennis, C., Bronowska, A., Paesen, G.C., Homans, S.W., 2010. Comparison of entropic contributions to binding in a "hydrophilic" versus "hydrophobic" ligand-protein interaction. J. Am. Chem. Soc. 132, 8682–8689.

Tellinghuisen, J., 2012. Designing isothermal titration calorimetry experiments for the study of 1:1 binding: problems with the "standard protocol". Anal. Biochem. 424, 211–220.

Tellinghuisen, J., Chodera, J.D., 2011. Systematic errors in isothermal titration calorimetry: concentrations and baselines. Anal. Biochem. 414, 297–299.

Velazquez-Campoy, A., Freire, E., 2005. ITC in the post-genomic era? Priceless. Biophys. Chem. 115, 115–124.

Valezques-Campoy, A., Freire, E., 2006. Isothermal titration calorimetry to determine association constants for high-affinity ligands. Nat. Protoc. 1, 186–191.

Wang, J., Berne, B.J., Friesner, R.A., 2011. Ligand binding to protein-binding pockets with wet and dry regions. Proc. Natl. Acad. Sci. U.S.A. 108, 1326–1330.

Weber, I.T., Agniswamy, J., 2009. HIV-1 protease: structural perspective on drug resistance. Viruses 1, 1110–1136.

Wermuth, C.G., 2003. Application of strategies for primary structure-activity relationship exploration. In: Wermuth, C.G. (Ed.), The Practice of Medicinal Chemistry. Elsevier (Chapter 18).

Young, T., Abel, R., Kim, B., Berne, B.J., Friesner, R.A., 2007. Motifs for molecular recognition exploiting hydrophobic enclosure in protein-ligand binding. Proc. Natl. Acad. Sci. U.S.A. 104, 808–813.

Zhang, Y.-L., Zhang, Z.-Y., 1998. Low-affinity binding determined by titration calorimetry using a high-affinity coupling ligand: a thermodynamic study of ligand binding to protein tyrosine phosphatase 1B. Anal. Biochem. 261, 139–148.

Zidek, L., Novotny, M.V., Stone, M.J., 1999. Increased protein backbone conformational entropy upon hydrophobic ligand binding. Nat. Struct. Biol. 6, 1118–1121.

Chapter 2

Machine Learning Approach to Predict Enzyme Subclasses

R. Concu[1], H. González-Díaz[2,3], M.N.D.S. Cordeiro[1]

[1]*University of Porto, Porto, Portugal;* [2]*University of the Basque Country UPV/EHU, Bilbao, Bizkaia, Spain;* [3]*IKERBASQUE, Basque Foundation for Science, Bilbao, Spain*

2.1. INTRODUCTION

Predicting drug–protein interaction is a prime goal for drug development (Shi et al., 2015; Peng et al., 2015). While in the past the main purpose in drug discovery was to develop highly selective drugs, nowadays researches are mainly focused on finding new targets for the just-discovered drugs (Shi et al., 2015; Ma et al., 2016). One of the most common ways of achieving this goal is to build up a network map representing the interaction between proteins (Yugandhar and Gromiha, 2016; Wang et al., 2016; Gable et al., 2016). In this context, predicting new enzymes or assigning an unknown enzyme to a specific subclass is an essential step in this process. Many authors have proposed different kinds of approaches to deal with this problem. In this chapter, we present a new strategy to predict enzyme subclasses using a lattice representation.

In the past years, several different graph representations have been extended to theoretical biology. In this context, many authors are using graph theory to represent the structure and/or dynamics of large biological systems such as protein–protein interaction networks. Complex networks consist of nodes and edges/arcs (node–node connections or links). Drugs, genes, RNAs, proteins, organisms, brain cortex regions, diseases, patients, or environmental systems may play the role of nodes, while the edges/connections represent similarity/dissimilarity relationships between the nodes. In complex networks, both nodes and edges are placed in the space without any kind of particular geometrical correlation; in fact, in general, complex network nodes do not need spatial coordinates, and edges do not have a specific length or shape (Boccaletti et al., 2006; Barabasi and Oltvai, 2004; Estrada, 2006). However, Randic, Nandy, Basak, Liao, and many others have developed and introduced in biology some special types of graph-based representations, including geometrical constraints to node positioning (sequence pseudofolding rules) in

Multi-Scale Approaches in Drug Discovery. http://dx.doi.org/10.1016/B978-0-08-101129-4.00002-3

2D space, thereby adopting final geometrical shapes that resemble lattice-like patterns. Lattice networks are very powerful tools that have been used in the past to represent biological sequences such as DNA and protein; however, their applications have been extended to a lot of other fields. Indeed, this technique has been used to create string pseudofolding lattice representations for any kind of string data. In several previous reports, we have discussed the applications of different graphs in Proteomics and other Biomedical Sciences (Gonzalez-Diaz et al., 2008a,b; Gonzalez-Diaz, 2009).

Several authors have used pseudofolding lattice hydrophobicity−polarity models to simulate polymer folding by optimizing the lattice structure and resembling the real folding (Berger and Leighton, 1998). However, we can choose notably simpler polymer chain pseudofolding rules to avoid optimization procedures and notably speed up the construction of the lattice. Toward this goal, useful graph representations of DNA, RNA, and/or protein sequences have been introduced by Gates (1986), Nandy (1996), (Leong and Morgenthaler (1995)), and Randic et al. (2001) based on 2D coordinate systems. We call these graph representations as polymer sequence pseudofolding lattice networks because they look like lattice structures, and in fact, we force a sequence to fold in a way that does not necessarily occur in nature. In this regard, a novel 2D-lattice representation for protein sequence similar to the one proposed by Nandy for DNA sequences was introduced by our group in the study of protein sequences (Nandy, 2003, 1996; Roy et al., 1998). In such 2D graph, each of the four amino acid groups is assigned to each axis direction according to the physicochemical nature of the amino acids, i.e., nonpolar and noncharged, polar but noncharged, positively charged, or negatively charged (Aguero-Chapin et al., 2006). These four classes characterize the physicochemical nature of the amino acids as polar, nonpolar, acid, or basic, respectively. Classification as positively or negatively charged prevails over polar/nonpolar classification in such a way that the four classes do not overlap each other. In mathematical terms, it means that, in this example, we used the vector $\mathbf{s} = [s_0, s_1, s_2, \ldots s_j, \ldots s_n]$ to list the labels for the n amino acids s_j, which are the elements of the system (protein sequence). Here we also used two vectors of weights to numerically characterize the s_j. The first vector ${}^1\mathbf{w} = [q_0, q_1, q_2, \ldots q_j, \ldots q_n]$ lists the electrostatic charge of each of the n amino acids in the sequence of the protein or peptide. The second vector ${}^2\mathbf{w} = [\mu_1, \mu_2, \mu_3, \ldots \mu_i, \ldots \mu_n]$ lists the dipolar moments of each amino acid. We also used herein the sets of conditions C1, C2, C3, and C4 that consist of two logical order operations. First, we place the node of the initial amino acid s_0 at the coordinates (0, 0) in a Cartesian 2D space. The coordinates of the successive amino acids are then calculated as follows in a manner similar to that for DNA spaces:

$\mathbf{C}_1$: Increase by +1 in the y axis if $q_j > 0$ (upwards-step); or
$\mathbf{C}_2$: Increase by +1 in the x axis if $q_j = 0$ and $\mu_j \neq 0$ (rightwards-step); or
$\mathbf{C}_3$: Decrease by −1 in the y axis if $q_j < 0$ (downwards-step); or
$\mathbf{C}_4$: Decrease by −1 in the x axis otherwise (leftwards-step).

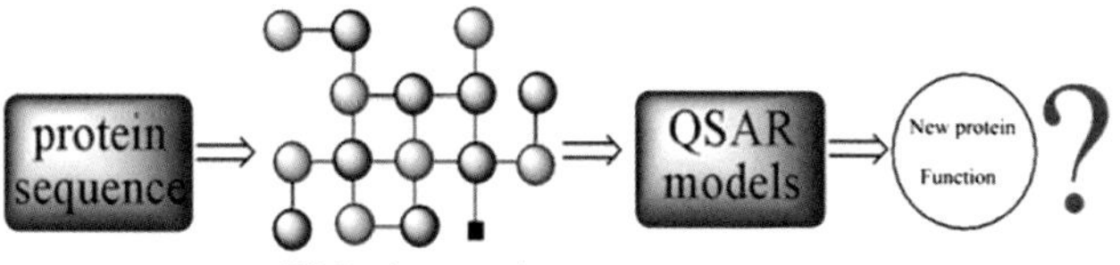

FIGURE 2.1 Flowchart of a pseudofolding QSAR study.

This new kind of representation is very similar to the those previously reported for DNA; however, in this new approach, we include a protein sequence of 20 amino acid types instead of a DNA sequence of four base types. The great advantage for the representation that we are proposing is that it overcomes the aforementioned 10D space bottleneck by previously grouping the 20 natural amino acids instead of making only four groups. Moreover, despite the proven efficacy of the new lattice-like graph/networks to represent diverse systems, past works focused only on one specific type of biological data. Due to this, in 2009 we have introduced a generalized type of lattice, illustrating how to use it to represent and compare biological data from different sources (Gonzalez-Diaz et al., 2009). Specifically, we extended the method from protein sequence to mass spectra (MS) of peptide mass fingerprints (PMF), molecular dynamics (MD) results from protein structural studies, mRNA microarray data, single nucleotide polymorphisms, 1D or 2D electrophoretic (2DE) study of protein polymorphisms, and protein research patent and/or copyright information. Thus, in a work reported in 2009, we developed a generalized type of lattice graphs useful to represent strings of biological information (Gonzalez-Diaz et al., 2009).

A main application of these lattices is the calculation of sequence pseudofolding parameters to construct quantitative structure—activity relationship, or in general, quantitative structure—property relationship, (QSAR or QSPR, respectively)—like models (Toropova et al., 2016a,b; Fioressi et al., 2015). Generally speaking, QSAR/QSPR-like models are capable of connecting the structure of a system with its implicit properties, which are not self-evident after direct inspection of the structure (Gonzalez-Diaz, 2008; González-Díaz et al., 2011, 2008a; González-Díaz and Munteanu, 2010). In Fig. 2.1, we illustrate the common pathway used in the development of QSAR/QSPR models using molecular descriptors derived from lattice graphs. The reported workflow is the most common in QSAR/QSPR studies.

2.2. MATERIAL AND METHODS

2.2.1. Background for Enzyme Subclasses Prediction

The isolation and prediction of new enzymes is the main goal for drug development, drug target prediction, and protein—protein interaction. A common strategy used to accomplish the former goal is the use of classical experimental techniques based on proteomics. For instance, in proteomic research, scientists often use a combination of 2DE and mass spectrometry (MS) to isolate and characterize new sequences from biological samples (Aksu et al., 2002). In these cases, we employ informatics tools such as the MASCOT search engine, to have the MS outcomes for some of the most important peptides of the most similar proteins (Hirosawa et al., 1993; Resing et al., 2004). In addition, one alternative is the application of alignment-free machine learning methods for predicting the protein's functional classes based on sequence (1D) parameters but independently of their sequence–sequence similarity (Lin et al., 2006a,b; Han et al., 2004, 2005). Thus, the so-called pseudo amino acid composition indices and other predictive methods have been introduced by Chou et al. and Cai et al. (Cai et al., 2005; Chou, 2001, 2005; Wang et al., 2006; Xiao et al., 2006; Cai and Chou, 2006, 2005a; Chou and Cai, 2003, 2005), and other authors to predict the function of proteins irrespective of their sequence similarity. Some authors developed new techniques to predict the Enzyme Commission (EC) (Nomenclature) number or the protein function with very good results, like the group of Cai et al. that have used the functional domain composition approach (Cai and Chou, 2005b). On the other hand, Dobson and Doig presented a model based on support vector machine and structural parameters derived from protein 3D structure (Dobson and Doig, 2005) to discriminate between enzymes and nonenzymes.

In this work we approach the problem with the method MARCH-INSIDE 2.0. In the case of proteins, MARCH-INSIDE 2.0 can be used to calculate not only different parameters for the protein 3D structure but also parameters of pseudofolding lattice graphs derived from protein sequences. In this work we use MARCH-INSIDE 2.0 to develop a new machine learning model that predicts the probability of a protein sequence to belong to one of the more than 60 different EC2 subclasses of enzymes derived from the 6 EC main classes.

2.2.2. Computational Model

2.2.2.1. Input Parameters

In prior works, we have predicted the protein's function based on the spectral moments θ_k values of lattice graphs. It should be noted that the spectral moments depend on the probability ${}^kP_{ij}$ with which the effect of one particular interaction propagates from amino acid i to other amino acids j close to it in the 2D pseudofolding space and return to ith amino acid after k steps. These spectral moments are straightforward to calculate as they are defined as the trace (Tr) of the kth power (${}^k\mathbf{\Pi}$) of the stochastic matrix ${}^1\mathbf{\Pi}$. In this chapter, we present a global model using spectral moments, entropy, and mean property molecular descriptors (md).

2.2.2.2. Data Set

The list of the proteins used in this study was compiled by downloading all the proteins belonging to one of the EC subclasses (Dobson and Doig, 2005).

The sequence of all proteins were downloaded in FASTA format from the Protein Data Bank (PDB) (Ivanisenko et al., 2005). The final data set used in this work contains 17,620 enzyme proteins and a total of 65,535 entries including nonenzyme proteins. These cases include proteins that do not belong to any of the 69 subclasses of enzymes. Moreover, since one protein may be repeated as a negative case for different subclasses, we included a number of negative cases on each subclass proportional to its number of positive cases. Consequently, we compiled a data set of 47,915 negative cases. This data set is notably larger than the one previously used by other authors or our group, which consisted of only 4755 proteins including enzymes and nonenzymes. Here, we adopt definitions within the EC framework. This scheme dates back to 1956, when the International Union of Biochemistry and Molecular Biology (IUBMB) began to regularize the naming and categorization of enzymes. In fact, the EC system has a hierarchical structure developed with six classes of enzymes at the top level determined by the general reaction catalyzed. They are oxidoreductases, transferases, hydrolases, lyases, isomerases, and ligases (Babbitt, 2003). In Table 2.1 we report the EC codes present in our database according to the first term of the EC number notation, the number of enzymes present in each subclass, and the classification accuracy of the models. Detailed information for each protein may be found in the Supporting Information (SI: Table SM1).

2.2.2.3. Multitarget QSAR Statistical Method

As regard the machine learning technique, we opted for using artificial neural networks (ANN) to build up the classifier. One of the most important steps in this work was the organization of the spreadsheet containing the raw data used as input for the ANN because this is not a classic classifier. A schematization on how to setup such type of spreadsheets has been previously described by our group (Gonzalez-Diaz et al., 2012). In any case, our aim is to gather a discriminant function able to classify proteins into two groups, that is, those that belong to a particular predefined group and those that do not belong. For doing so, we have to indicate somehow what group we pretend to predict in each case. Toward this end, we followed the steps below:

1. First, the proteins belonging to the group of enzymes were divided according to their subclass membership using the nomenclature EC number of level 2 (Nomenclature). The family of enzymes is the first number of the aforementioned nomenclature, and the subclass is the second number in this classification. This information is contained in the protein's PDB file (Ivanisenko et al., 2005).
2. Then, we created raw data representing each protein input as a vector composed of one output variable, six structural variables (inputs), and the enzyme class query (ECQ) variable. ECQ is an auxiliary variable not used to construct the model. We also included six average values relative to the subclass and six deviations of the values of the protein from the subclass

TABLE 2.1 Enzyme Subclasses, Functions, and Classification Accuracy for all the Subclasses

Subclass	Function	Number	Correct	Incorrect	%
EC 1.1	Acting on the CH–OH group of donors	560	516	44	95.89
EC 1.2	Acting on the aldehyde or oxo group of donors	229	184	45	93.01
EC 1.3	Acting on the CH–CH group of donors	183	157	26	91.80
EC 1.4	Acting on the CH–NH_2 group of donors	120	100	20	85.00
EC 1.5	Acting on the CH–NH group of donors	122	98	24	85.25
EC 1.6	Acting on NADH or NADPH	122	43	79	67.21
EC 1.7	Acting on other nitrogenous compounds as donors	92	66	26	83.70
EC 1.8	Acting on a sulfur group of donors	148	87	61	77.70
EC 1.9	Acting on a heme group of donors	70	40	30	61.43
EC 1.10	Acting on diphenols and related substances as donors	201	100	101	49.25
EC 1.11	Acting on a peroxide as acceptor	131	107	24	86.26
EC 1.13	Acting on single donors with incorporation of molecular oxygen (oxygenases)	120	98	22	82.50
EC 1.14	Acting on paired donors, with incorporation or reduction of molecular oxygen	254	202	52	82.28
EC 1.15	Acting on superoxide radicals as acceptor	170	115	55	74.12
EC 1.16	Oxidizing metal ions	178	136	42	93.26

TABLE 2.1 Enzyme Subclasses, Functions, and Classification Accuracy for all the Subclasses—cont'd

Subclass	Function	Number	Correct	Incorrect	%
EC 1.17	Acting on CH or CH_2 groups	82	66	16	81.71
EC 1.18	Acting on iron-sulfur proteins as donors	66	50	16	77.27
EC 1.21	Acting on the reaction X–H + Y–H = X–Y	27	27	0	100.00
EC 1.97	Other oxidoreductases	80	75	5	93.75
EC 2	Transferases				
EC 2.1	Transferring one-carbon groups	565	494	71	89.38
EC 2.2	Transferring aldehyde or ketonic groups	82	76	6	91.46
EC 2.3	Acyltransferases	577	466	111	84.40
EC 2.4	Glycosyltransferases	447	360	87	86.35
EC 2.5	Transferring alkyl or aryl groups, other than methyl groups	279	226	53	80.29
EC 2.6	Transferring nitrogenous groups	296	275	21	91.89
EC 2.7	Transferring phosphorus-containing groups	3222	3010	212	93.95
EC 2.8	Transferring sulfur-containing groups	65	52	13	87.69
EC 3	Hydrolases				
EC 3.1	Acting on ester bonds	1870	1688	182	89.36
EC 3.2	Glycosylases	745	656	89	90.34
EC 3.3	Acting on ether bonds	169	158	11	94.08
EC 3.4	Acting on peptide bonds (peptidases)	1543	1360	183	86.78
EC 3.5	Acting on carbon-nitrogen bonds, other than peptide bonds	570	470	100	84.04
EC 3.6	Acting on acid anhydrides	754	592	162	84.35

Continued

TABLE 2.1 Enzyme Subclasses, Functions, and Classification Accuracy for all the Subclasses—cont'd

Subclass	Function	Number	Correct	Incorrect	%
EC 3.7	Acting on carbon–carbon bonds	69	59	10	91.30
EC 3.8	Acting on halide bonds	86	64	22	77.91
EC 4	Lyases				
EC 4.1	Carbon–carbon lyases	580	485	95	87.76
EC 4.2	Carbon-oxygen lyases	540	479	61	90.74
EC 4.3	Carbon-nitrogen lyases	84	62	22	85.71
EC 4.6	Phosphorus-oxygen lyases	79	63	16	83.54
EC 4.99	Other lyases	100	86	14	85.00
EC 5	Isomerases				
EC 5.1	Racemases and epimerases	187	133	54	77.01
EC 5.2	Cis-trans-isomerases	72	44	28	47.22
EC 5.3	Intramolecular isomerases	258	192	66	74.81
EC 5.4	Intramolecular transferases (mutases)	149	134	15	85.91
EC 5.5	Intramolecular lyases	131	115	16	87.79
EC 5.99	Other isomerases	108	93	15	96.30
EC 6	Ligases				
EC 6.1	Forming carbon—oxygen bonds	306	259	47	88.24
EC 6.2	Forming carbon—sulfur bonds	27	22	5	88.89
EC 6.3	Forming carbon—nitrogen bonds	423	319	104	72.58
EC 6.4	Forming carbon–carbon bonds	137	105	1	97.17

average. Thus, the input vector for each protein contains 18 structural variables in total.

3. The first element (output) is a dummy variable (Boolean) called observed group (OG); OG = 1 if the protein belongs to the subclass we refer in ECQ and 0 otherwise (OG = 0).
4. The problem in this type of organization for raw data is that the six π_k values are protein constants. Consequently, the application of linear discriminant analysis based only on these values will fail necessarily when we change the OG values. A problem in this regard is that if we pretend to use the model for a real enzyme we have only one unspecific prediction and we need up to 69 specific probabilities, that is, 1 confirming the real subclass and 68 giving low probabilities for the other 68 subclasses. One can, however, solve this problem by introducing characteristic variables for each EC subclass referred on the ECQ but without having to provide information in the input about the real EC subclass of the protein. To do so, we used the average value of each π_k for all enzymes that belong to the same EC subclass. We also calculated the deviation of the ${}^{dev}\pi_k$ from the respective group indicated in ECQ. It is important to understand that we never used as input ECQ, so the model only includes as input the values for the protein entry and the average and deviations of these values from the ECQ, that is, not necessarily the real EC class.

Variable selection was carried out by the STATISTICA program that was used to build up the models. All the variables included in the model were standardized to bring them onto the same scale.

2.3. RESULTS

The search for approaches that complement or improve classical alignment tools for protein function annotation, like BLAST, continues to be the main goal; see for instance the works of Ivanciuc (Ivanciuc, 2008; Schein et al., 2007; Ivanciuc and Braun, 2007; Ivanciuc et al., 2002, 2004). In particular, predicting the EC class of one protein is highly relevant for developing future applications for enzyme design (Song et al., 2006; Jones et al., 2005; Zehetner, 2003; Yang, 2002; Lee et al., 2002). In a previous work, we accomplished this goal through a linear 2D-QSAR study (Gonzalez-Diaz et al., 2012). In that work, we developed a series of complex ANN models to assign each protein to one of six enzyme classes or predict it as nonenzyme using as inputs the 3D potential parameters. The five best models found are reported in Table 2.2.

As one can see, all the reported models have a good overall classification of enzymes and nonenzymes with a consistent accuracy in both training and validation series. In addition, all the models are able to give a good classification for the subclasses. Looking at Table 2.2, one can notice and assure that the best model is the one numbered 4, thus from now on we will report in more detail its results.

TABLE 2.2 Training and Validation Results for the Five Best Models Found

Index	Net. Name	Training	Validation	Training Algorithm	Error Function	Hidden Activation	Output Activation
1	MLP 54-38-2	93.13	92.88	BFGS 411	Entropy	Logistic	Softmax
2	MLP 54-50-2	93.76	93.44	BFGS 392	Entropy	Tanh	Softmax
3	MLP 54-50-2	92.96	92.13	BFGS 363	Entropy	Tanh	Softmax
4	MLP 54-50-2	94.35	93.87	BFGS 453	Entropy	Tanh	Softmax
5	MLP 54-50-2	93.54	92.58	BFGS 424	Entropy	Tanh	Softmax

In the training series, 12,257 enzymes were correctly classified and 1736 were misclassified (sensitivity = 87.00%) and 31,305 nonenzymes were correctly classified and 868 were misclassified (specificity = 97.00%). In the validation series, 2980 enzymes were correctly classified and 471 misclassified (sensitivity = 86.00%), 7854 nonenzymes were correctly classified and 236 misclassified (specificity = 97.00%). All the statistical data of this model are resumed in Table 2.3, while in Table 2.1 we show the classification accuracy for all the subclasses included in the model. By considering these results, one can see that the level of accuracy shown is adequate and in line with other machine learning methods developed to predict enzymes subclasses (Syed and Yona, 2009; Kristensen et al., 2008; Espadaler et al., 2008; Astikainen et al., 2008; Zhou et al., 2007; Huang et al., 2007). Moreover, this is also confirmed by looking at the Matthews's correlation coefficient (MCC = 0.86; Matthews, 1975) obtained for this model. In Fig. 2.2, the receiver operating characteristic

TABLE 2.3 Training and Validation Accuracy for the Best Model Found

Parameters	Group	%	Nonenzymes	Enzymes
Training				
Specificity	Nonenzymes	97.00	31,305	868
Sensitivity	Enzymes	87.00	1736	12,257
Accuracy	Total	94.36		
Validation				
Specificity	Nonenzymes	97.00	7854	236
Sensitivity	Enzymes	86.00	471	2980
Accuracy	Total	93.87		

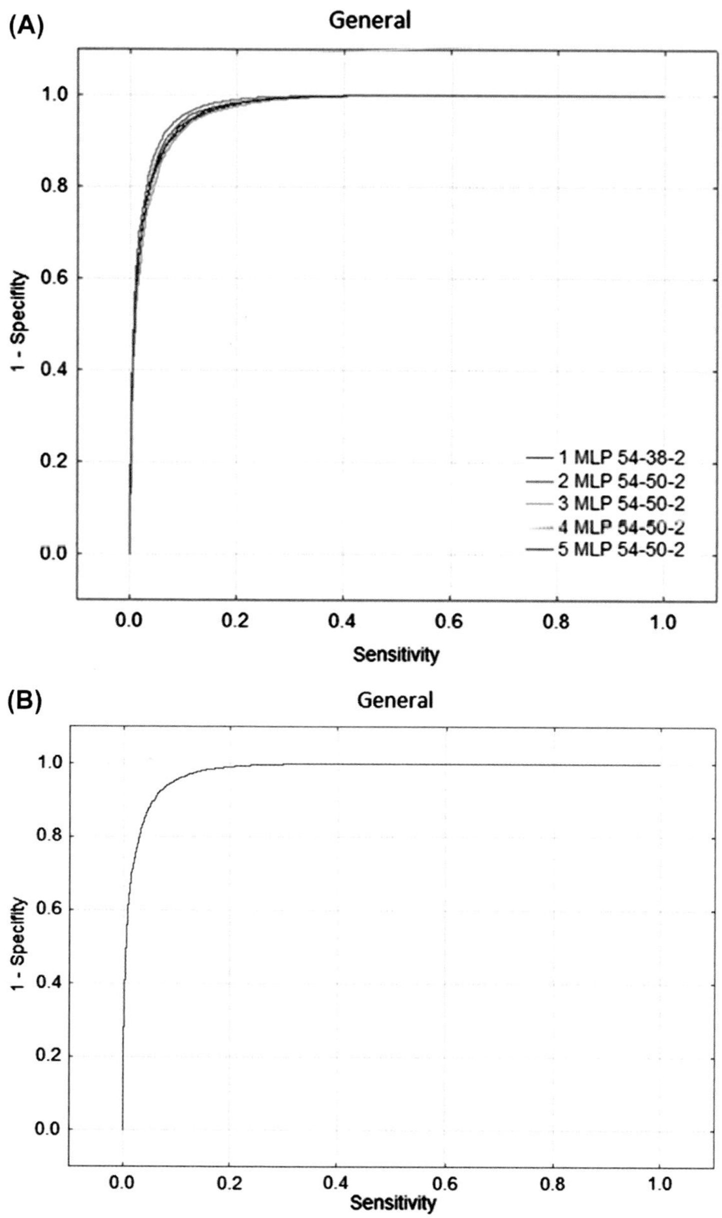

FIGURE 2.2 Receiver operating characteristic curve of all the models (A), and of the best one (B, C, D).

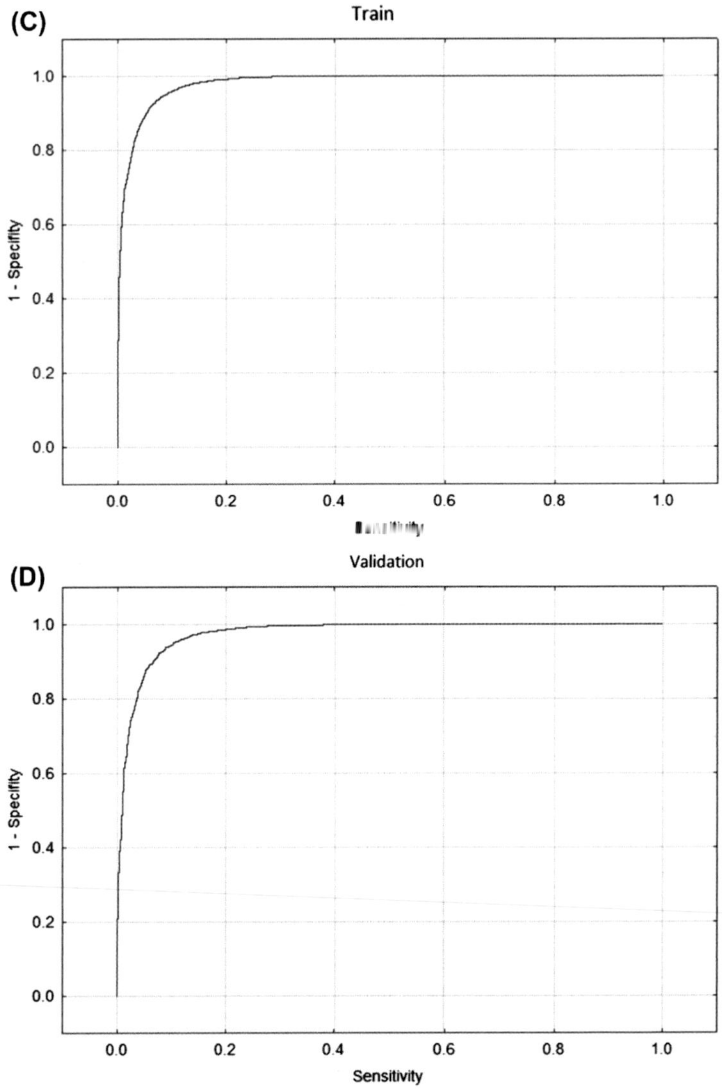

FIGURE 2.2 cont'd

(ROC) curves attained for the five best models are displayed. Table SM2 of SI shows in addition the classification accuracy of the five models.

2.4. DISCUSSION

Predicting enzyme subclasses is a challenging task due to various problems. First, the wide variety in enzymes regarding size, shape, amino acid composition, etc., greatly increases the number of molecular descriptors needed. This means that the chemical space covered by such a big data set like ours may be really wide, and finding a good correlation between the inputs (md) and outputs (EC subclasses) might be a really difficult task. Due to this, it is essential to choose a well-studied technique to find a robust model. We choose to develop our models using ANN in combination with the multilayer perceptron algorithm because over the years the latter has demonstrated its prediction ability in different fields (Zhang et al., 2015; Pourahmad et al., 2015; Cheng et al., 2008; Rossi and Conan-Guez, 2005). As one can notice from Table 2.2, all the models have a good balance between the accuracy values in the training and validation series. This means that the models are able to extract from the inputs the essential information forward reliably predicting unknown enzymes. Regarding the best model, one can see that there is a slight difference between the accuracy for the positive and negatives series. This trend is further confirmed in the other models. In our opinion, this difference is due to the difference in the entries between the series, for instance, 17,444 against 40,263. Even so, the robustness of the best model (and also of the others) is further confirmed by both the corresponding MCC value (=0.86) and ROC curve. Analyzing the classification accuracy of each subclass, it can be seen how the model is able to correctly classify almost all the subclasses save for the classes 1.10, 1.6, 1.9, and 5.2. Regarding the topology of the ANN models, it can be deduced that except the first model, which uses a layer of 38 neurons, all the others have a layer of 50 neurons. One might judge that those neurons' numbers are too high, but indeed they are essential to find good correlations and thus, robust models.

2.5. CONCLUSIONS

The setup of 2D-generalized lattice graphs constrained into a Cartesian coordinate system is a useful technique for biological data visualization, which in turn is not necessarily limited to DNA sequences. For instance, in this work, we have shown how it is possible to extend this methodology for classifying enzyme subclasses. Moreover, the usefulness of applying machine learning approaches is clearly grounded by the results attained and here presented toward our main goal, that is, be able to effectively predict enzyme subclasses.

ACKNOWLEDGMENTS

This work received financial support from Fundação para a Ciência e a Tecnologia (FCT/MEC) through national funds, and was cofinanced by the European Union (FEDER funds) under the Partnership Agreement PT2020, through projects UID/QUI/50006/2013, POCI/01/0145/FEDER/007265, and NORTE-01-0145-FEDER-000011 (LAQV@REQUIMTE). RC also acknowledges FCT and the European Social Fund for financial support (Grant SFRH/BPD/80605/2011). The authors are greatly indebted to all financing sources.

REFERENCES

Aguero-Chapin, G., Gonzalez-Diaz, H., Molina, R., Varona-Santos, J., Uriarte, E., Gonzalez-Diaz, Y., 2006. Novel 2D maps and coupling numbers for protein sequences. The first QSAR study of polygalacturonases; isolation and prediction of a novel sequence from *Psidium guajava* L. FEBS Lett. 580 (3), 723–730.

Aksu, S., Scheler, C., Focks, N., Leenders, F., Theuring, F., Salnikow, J., Jungblut, P.R., 2002. An iterative calibration method with prediction of post-translational modifications for the construction of a two-dimensional electrophoresis database of mouse mammary gland proteins. Proteomics 2 (10), 1452–1463.

Astikainen, K., Holm, L., Pitkanen, E., Szedmak, S., Rousu, J., 2008. Towards structured output prediction of enzyme function. BMC Proc. 2 (Suppl. 4), S2.

Babbitt, P.C., 2003. Definitions of enzyme function for the structural genomics era. Curr. Opin. Chem. Biol. 7, 230–237.

Barabasi, A.L., Oltvai, Z.N., 2004. Network biology: understanding the cell's functional organization. Nat. Rev. Genet. 5 (2), 101–113.

Berger, B., Leighton, T., 1998. Protein folding in the hydrophobic-hydrophilic (HP) model is NP-complete. J. Comput. Biol. 5 (1), 27–40.

Boccaletti, S., Latora, V., Moreno, Y., Chavez, M., Hwang, D.U., 2006. Complex networks: structure and dynamics. Phys. Rep. 424, 175–308.

Cai, Y.D., Chou, K.C., 2005a. Predicting membrane protein type by functional domain composition and pseudo-amino acid composition. J. Theor. Biol.

Cai, Y.D., Chou, K.C., 2005b. Using functional domain composition to predict enzyme family classes. J. Proteome Res. 4 (1), 109–111.

Cai, Y.D., Chou, K.C., 2006. Predicting membrane protein type by functional domain composition and pseudo-amino acid composition. J. Theor. Biol. 238 (2), 395–400.

Cai, Y.D., Zhou, G.P., Chou, K.C., 2005. Predicting enzyme family classes by hybridizing gene product composition and pseudo-amino acid composition. J. Theor. Biol. 234 (1), 145–149.

Cheng, J., Xiao, Q., Li, X.W., Liu, Q.H., Du, Y.M., 2008. Multi-layer perceptron neural network based algorithm for simultaneous retrieving temperature and emissivity from hyperspectral FTIR data. Guang Pu Xue Yu Guang Pu Fen Xi 28 (4), 780–783.

Chou, K.C., Cai, Y.D., 2003. Predicting protein quaternary structure by pseudo amino acid composition. Proteins 53 (2), 282–289.

Chou, K.C., Cai, Y.D., 2005. Prediction of membrane protein types by incorporating amphipathic effects. J. Chem. Inf. Model. 45 (2), 407–413.

Chou, K.C., 2001. Prediction of protein cellular attributes using pseudo-amino acid composition. Proteins 43 (3), 246–255 (Erratum: ibid., 2001, vol. 44, 60).

Chou, K.C., 2005. Using amphiphilic pseudo amino acid composition to predict enzyme subfamily classes. Bioinformatics 21 (1), 10–19.

Dobson, P.D., Doig, A.J., 2005. Predicting enzyme class from protein structure without alignments. J. Mol. Biol. 345 (1), 187–199.

Espadaler, J., Eswar, N., Querol, E., Aviles, F.X., Sali, A., Marti-Renom, M.A., Oliva, B., 2008. Prediction of enzyme function by combining sequence similarity and protein interactions. BMC Bioinform. 9, 249.

Estrada, E., 2006. Protein bipartivity and essentiality in the yeast protein-protein interaction network. J. Proteome Res. 5 (9), 2177–2184.

Fioressi, S.E., Bacelo, D.E., Cui, W.P., Saavedra, L.M., Duchowicz, P.R., 2015. QSPR study on refractive indices of solvents commonly used in polymer chemistry using flexible molecular descriptors. SAR QSAR Environ. Res. 26 (6), 499–506.

Gable, J.E., Lee, G.M., Acker, T.M., Hulce, K.R., Gonzalez, E.R., Schweigler, P., Melkko, S., Farady, C.J., Craik, C.S., 2016. Fragment-based protein–protein interaction antagonists of a viral dimeric protease. ChemMedChem 11 (8), 862–869.

Gates, M.A., 1986. A simple way to look at DNA. J. Theor. Biol. 119, 319–328.

González-Díaz, H., Munteanu, C.R., 2010. Topological Indices for Medicinal Chemistry, Biology, Parasitology, Neurological and Social Networks. Transworld Research Network, Kerala, India, pp. 001–212.

Gonzalez-Diaz, H., Gonzalez-Diaz, Y., Santana, L., Ubeira, F.M., Uriarte, E., 2008a. Proteomics, networks and connectivity indices. Proteomics 8 (4), 750–778.

Gonzalez-Diaz, H., Prado-Prado, F., Ubeira, F.M., 2008b. Predicting antimicrobial drugs and targets with the MARCH-INSIDE approach. Curr. Top. Med. Chem. 8 (18), 1676–1690.

Gonzalez-Diaz, H., Perez-Montoto, L.G., Duardo-Sanchez, A., Paniagua, E., Vazquez-Prieto, S., Vilas, R., Dea-Ayuela, M.A., Bolas-Fernandez, F., Munteanu, C.R., Dorado, J., Costas, J., Ubeira, F.M., 2009. Generalized lattice graphs for 2D-visualization of biological information. J. Theor. Biol. 261 (1), 136–147.

González-Díaz, H., Prado-Prado, F., García-Mera, X., 2011. Complex Network Entropy: From Molecules to Biology, Parasitology, Technology, Social, Legal, and Neurosciences. Transworld Research Network, Kerala, India, pp. 001–142.

Gonzalez-Diaz, H., C, R., Perez-Montoto, L.G., Ubeira, F.M., Romaris, F., Paniagua, E., Duardo-Sanchez, A., Prado-Prado, F., 2012. Generalized string pseudo-folding lattices in bioinformatics: state-of-art review, new model for enzyme sub-classes, and study of ESTs on *Trichinella spiralis*. Curr. Bioinform. 7 (1). See more at: http://www.eurekaselect.com/89589/article#sthash.NAEFkypH.dpuf.

Gonzalez-Diaz, H., 2008. Quantitative studies on structure–activity and structure–property relationships (QSAR/QSPR). Curr. Top. Med. Chem. 8 (18), 1554.

Han, L.Y., Cai, C.Z., Ji, Z.L., Cao, Z.W., Cui, J., Chen, Y.Z., 2004. Predicting functional family of novel enzymes irrespective of sequence similarity: a statistical learning approach. Nucleic Acids Res. 32 (21), 6437–6444.

Han, L.Y., Cai, C.Z., Ji, Z.L., Chen, Y.Z., 2005. Prediction of functional class of novel viral proteins by a statistical learning method irrespective of sequence similarity. Virology 331 (1), 136–143.

Hirosawa, M., Hoshida, M., Ishikawa, M., Toya, T., 1993. MASCOT: multiple alignment system for protein sequences based on three-way dynamic programming. Comput. Appl. Biosci. 9 (2), 161–167.

Huang, W.L., Chen, H.M., Hwang, S.F., Ho, S.Y., 2007. Accurate prediction of enzyme subfamily class using an adaptive fuzzy *k*-nearest neighbor method. Biosystems 90 (2), 405–413.

Ivanciuc, O., Braun, W., 2007. Robust quantitative modeling of peptide binding affinities for MHC molecules using physical-chemical descriptors. Protein Pept. Lett. 14 (9), 903–916.

Ivanciuc, O., Schein, C.H., Braun, W., 2002. Data mining of sequences and 3D structures of allergenic proteins. Bioinformatics 18 (10), 1358–1364.

Ivanciuc, O., Oezguen, N., Mathura, V.S., Schein, C.H., Xu, Y., Braun, W., 2004. Using property based sequence motifs and 3D modeling to determine structure and functional regions of proteins. Curr. Med. Chem. 11 (5), 583–593.

Ivanciuc, O., 2008. Weka machine learning for predicting the phospholipidosis inducing potential. Curr. Top. Med. Chem. 8 (18), 1691–1709.

Ivanisenko, V.A., Pintus, S.S., Grigorovich, D.A., Kolchanov, N.A., 2005. PDBSite: a database of the 3D structure of protein functional sites. Nucleic Acids Res. 33 (Database issue), D183–D187.

Jones, C.E., Baumann, U., Brown, A.L., 2005. Automated methods of predicting the function of biological sequences using GO and BLAST. BMC Bioinform. 6, 272.

Kristensen, D.M., Ward, R.M., Lisewski, A.M., Erdin, S., Chen, B.Y., Fofanov, V.Y., Kimmel, M., Kavraki, L.E., Lichtarge, O., 2008. Prediction of enzyme function based on 3D templates of evolutionarily important amino acids. BMC Bioinform. 9, 17.

Lee, C., Grasso, C., Sharlow, M.F., 2002. Multiple sequence alignment using partial order graphs. Bioinformatics 18 (3), 452–464.

Leong, P.M., Morgenthaler, S., 1995. Random walk and gap plots of DNA sequences. Comput. Applic Biosci. 11, 503–507.

Lin, H.H., Han, L.Y., Cai, C.Z., Ji, Z.L., Chen, Y.Z., 2006a. Prediction of transporter family from protein sequence by support vector machine approach. Proteins 62 (1), 218–231.

Lin, H.H., Han, L.Y., Zhang, H.L., Zheng, C.J., Xie, B., Chen, Y.Z., 2006b. Prediction of the functional class of lipid binding proteins from sequence-derived properties irrespective of sequence similarity. J. Lipid Res. 47 (4), 824–831.

Mu, W., Hu, L., Zhang, N., Dong, X., Li, Y., Yang, B., Tian, W., Wang, X., 2016. Prediction of candidate drugs for treating pancreatic cancer by using a combined approach. PLoS One 11 (2), e0149896.

Matthews, B.W., 1975. Comparison of the predicted and observed secondary structure of T4 phage lysozyme. Biochim. Biophys. Acta 405 (2), 442–451.

Nandy, A., 1996. Two-dimensional graphical representation of DNA sequences and intron-exon discrimination in intron-rich sequences. Comput. Appl. Biosci. 12 (1), 55–62.

Nandy, A., 2003. Novel method for discrimination of conserved genes through numerical characterization of DNA sequences. Int. Electron. J. Mol. Des. 2, 000.

Nomenclature, C. o. B., Enzyme Nomenclature.

Peng, L., Liao, B., Zhu, W., Li, K., 2015. Predicting drug-target interactions with multi-information fusion. IEEE J. Biomed. Health Inform.

Pourahmad, S., Azad, M., Paydar, S., 2015. Diagnosis of malignancy in thyroid tumors by multi-layer perceptron neural networks with different batch learning algorithms. Glob. J. Health Sci. 7 (6), 46–54.

Randic, M., Guo, X., Basak, S.C., 2001. On the characterization of DNA primary sequences by triplet of nucleic acid bases. J. Chem. Inf. Comput. Sci. 41 (3), 619–626.

Resing, K.A., Meyer-Arendt, K., Mendoza, A.M., Aveline-Wolf, L.D., Jonscher, K.R., Pierce, K.G., Old, W.M., Cheung, H.T., Russell, S., Wattawa, J.L., Goehle, G.R., Knight, R.D., Ahn, N.G., 2004. Improving reproducibility and sensitivity in identifying human proteins by shotgun proteomics. Anal. Chem. 76 (13), 3556–3568.

Rossi, F., Conan-Guez, B., 2005. Functional multi-layer perceptron: a non-linear tool for functional data analysis. Neural Networks 18 (1), 45–60.

Roy, A., Raychaudhur, C., Nandy, A., 1998. Novel techniques of graphical representation and analysis of DNA sequences – a review. J. Biosci. 23 (1), 55–71.

Schein, C.H., Ivanciuc, O., Braun, W., 2007. Bioinformatics approaches to classifying allergens and predicting cross-reactivity. Immunol. Allergy Clin. North Am. 27 (1), 1–27.

Shi, J.Y., Yiu, S.M., Li, Y., Leung, H.C., Chin, F.Y., 2015. Predicting drug-target interaction for new drugs using enhanced similarity measures and super-target clustering. Methods 83, 98–104.

Song, J., Burrage, K., Yuan, Z., Huber, T., 2006. Prediction of *cis/trans* isomerization in proteins using PSI-BLAST profiles and secondary structure information. BMC Bioinform. 7 (1), 124.

Syed, U., Yona, G., 2009. Enzyme function prediction with interpretable models. Methods Mol. Biol. 541, 373–420.

Toropova, A.P., Schultz, T.W., Toropov, A.A., 2016a. Building up a QSAR model for toxicity toward *Tetrahymena pyriformis* by the Monte Carlo method: a case of benzene derivatives. Environ. Toxicol. Pharmacol. 42, 135–145.

Toropova, A.P., Toropov, A.A., Veselinovic, A.M., Veselinovic, J.B., Leszczynska, D., Leszczynski, J., 2016b. Monte Carlo based QSAR models for toxicity of organic chemicals to *Daphnia magna*. Environ. Toxicol. Chem. SETAC.

Wang, S.Q., Yang, J., Chou, K.C., 2006. Using stacked generalization to predict membrane protein types based on pseudo-amino acid composition. J. Theor. Biol.

Wang, Y.C., Chen, S.L., Deng, N.Y., Wang, Y., 2016. Computational probing protein–protein interactions targeting small molecules. Bioinformatics 32 (2), 226–234.

Xiao, X., Shao, S., Ding, Y., Huang, Z., Chou, K.C., 2006. Using cellular automata images and pseudo amino acid composition to predict protein subcellular location. Amino Acids 30 (1), 49–54.

Yang, A.S., 2002. Structure-dependent sequence alignment for remotely related proteins. Bioinformatics 18 (12), 1658–1665.

Yugandhar, K., Gromiha, M.M., 2016. Analysis of protein–protein interaction networks based on binding affinity. Curr. Protein Pept. Sci. 17 (1), 72–81.

Zehetner, G., 2003. OntoBlast function: from sequence similarities directly to potential functional annotations by ontology terms. Nucleic Acids Res. 31 (13), 3799–3803.

Zhang, H., Gao, Y., Yuan, C., Liu, Y., Ding, Y., 2015. Research on early identification of bipolar disorder based on multi-layer perceptron neural network. J. Biom. Eng. 32 (3), 537–541.

Zhou, X.B., Chen, C., Li, Z.C., Zou, X.Y., 2007. Using Chou's amphiphilic pseudo-amino acid composition and support vector machine for prediction of enzyme subfamily classes. J. Theor. Biol. 248 (3), 546–551.

Chapter 3

Multitasking Model for Computer-Aided Design and Virtual Screening of Compounds With High Anti-HIV Activity and Desirable ADMET Properties

V.V. Kleandrova[1], A. Speck-Planche[2]
[1]Moscow State University of Food Production, Moscow, Russia; [2]University of Porto, Porto, Portugal

3.1. INTRODUCTION

Infectious agents have plagued mankind for centuries. Among them, the human immunodeficiency virus (HIV) is responsible for causing a severe infection, and ultimately, the life-threatening medical condition known as acquired immune deficiency syndrome (AIDS) (Fauci, 2008), which has become a global pandemic in the modern era. In 2012, there were approximately 35.3 million people living with HIV worldwide (Maartens et al., 2014), while in 2013, around 1.3 million AIDS deaths were estimated to have occurred in the top 30 countries with the highest AIDS mortality burden, representing 87% of global AIDS deaths (Granich et al., 2015). Antiretroviral regimens have been used in the fight against HIV/AIDS, and they have been so successful that in 2010, around 700,000 lives were saved by antiretroviral therapy alone (Fauci and Folkers, 2012). Despite these encouraging results, two major factors still represent great concerns: the emergence of multidrug resistance (Lima et al., 2008; Lohse et al., 2007; Napravnik et al., 2007; Tozzi et al., 2006) and the prevalence of serious side effects associated with the anti-HIV drugs (Luther and Glesby, 2007; Montessori et al., 2004; Nolan et al., 2005; Torres and Lewis, 2014). Therefore, the discovery of potent and safe anti-HIV drugs remains a challenging task for the scientific community.

Multi-Scale Approaches in Drug Discovery. http://dx.doi.org/10.1016/B978-0-08-101129-4.00003-5

Progresses in HIV research have been characterized by an explosion of experimental data, which have been stored, organized, and curated by different online repositories, CHEMBL database being the most significant of all (Gaulton et al., 2012; Mok and Brenk, 2011; Overington, 2009). In this context, chemoinformatic models based on quantitative structure–activity relationships (QSAR) have been used as tools to perform virtual screening of chemical libraries, contributing to gain more knowledge in early drug discovery, accelerating the search for highly active inhibitors against HIV-related proteins (Afantitis et al., 2006; Arkan et al., 2010; Bak and Polanski, 2006; Barreiro et al., 2007; Boutton et al., 2005; Fujii et al., 2003; Gupta et al., 2009; Niedbala et al., 2006; Perez-Nueno et al., 2008; Vedani et al., 2005; Waller et al., 1993), and rationalizing the serendipitous nature associated with organic synthesis and biological testing. Nevertheless, most of these prior scientific works have two major drawbacks. First, these QSAR models use small datasets of structurally related molecules. On the other hand, only one target (protein) has been considered in these studies.

Several researchers have emphasized the development of multitarget (mt) QSAR models, which have been able to predict pharmacological activity against many different biological targets (proteins, microorganisms, and cell lines) (Gonzalez-Diaz et al., 2006; Prado-Prado et al., 2007, 2008). In fact, several insightful works have been reported in HIV research (Prado-Prado et al., 2009; Speck-Planche and Kleandrova, 2012; Speck-Planche et al., 2012). Nevertheless, even with the remarkable progresses achieved by the mt-QSAR models, certain unfavorable issues prevail. For instance, in mt-QSAR models (as well as other QSAR models), only one measure of biological effect (activity, toxicity, etc.) is considered, and there is no information regarding the reliability of the assays under which the molecules are tested.

To overcome these issues, several types of advanced chemoinformatic models have emerged with the ability of performing simultaneous predictions of multiple biological effects of chemicals against many targets, and where the accuracy of the experimental information is also considered (Romero-Duran et al., 2016; Speck-Planche and Cordeiro, 2014, 2015; Speck-Planche et al., 2016; Tenorio-Borroto et al., 2014). Some of these advanced chemoinformatic models have involved the use of perturbation theory to simultaneously predict anti-HIV activities and AIDS prevalence in US counties. These are the first attempts to unify and link preclinical results with epidemiological and socioeconomic data (Gonzalez-Diaz et al., 2014; Herrera-Ibata et al., 2014, 2015). In any case, the creation of a model able to predict anti-HIV activities and safety profiles such as absorption, distribution, metabolism, elimination, and toxicity (ADMET) remains a prime goal in HIV research because it would open the horizons toward the development of more efficacious anti-HIV therapies. In this chapter, we introduce the first multitasking model for quantitative structure–biological effect relationships (mtk-QSBER), which is devoted to performing simultaneous prediction of anti-HIV activities and

ADMET profiles. Particularly, the mtk-QSBER model presented here is also focused on the fragment-based design and screening of new molecular entities with potential anti-HIV activity and desirable ADMET properties.

3.2. MATERIALS AND METHODS

3.2.1. Creation of the Data Set and Calculation of the Molecular Descriptors

The philosophy underpinning the construction of mtk-QSBER models has been explained in detail in previous works (Speck-Planche and Cordeiro, 2014, 2015). Therefore, only the main aspects will be mentioned. The data set containing the experimental data was extracted from CHEMBL (Gaulton et al., 2012; Mok and Brenk, 2011; Overington, 2009), and it was formed by 20,562 molecules. Each of them was assayed by considering at least 1 of 27 measures of biological effect (m_e), against at least 1 of 79 specific targets (b_t). At the same time, each experiment contained 1 of 4 labels of assay information (a_i) indicating whether the assay was focused on the study of binding phenomena; functional/physiological responses; ADME processes; or toxicological profiles. In addition, the assays were associated with at least 1 of 6 different target mappings (t_m), which indicated the general types of targets against which the biological tests were performed. Each combination of the elements m_e, b_t, a_i, and t_m defines a specific experimental condition, which can be expressed as an ontology of the form $c_j \rightarrow (m_e, b_t, a_i, t_m)$. This ontology was influenced by an external probabilistic factor (p_c) containing the reliability of the assay (experimental condition). Thus, in an increasing order, p_c took values of 0.55, 0.75, and 1 for the levels of experimental reliability named autocuration, intermediate, and expert, respectively.

All the molecules were not tested against all the experimental conditions, and therefore, the data set was formed by 29,682 cases. Each case in the data set was assigned to one of two possible classes/categories, namely, positive [$BE_i(c_j) = 1$, indicating potent anti-HIV activity or desirable ADMET property] or negative [$BE_i(c_j) = -1$, referred to weak anti-HIV activity or undesirable ADMET property]. All the assignments were realized according to certain cutoff values, which are represented in Table 3.1. It should be pointed out that $BE_i(c_j)$ is a binary variable that characterizes the biological effect of the ith molecule under the experimental condition c_j.

The SMILES codes for all the 29,682 cases were stored in a file of type *.txt, which was manually changed to *.smi. Afterward, the program Standardizer was used to convert the *.smi file to *.sdf (ChemAxon, 1998–2016). Then, from the *.sdf file, the program QUBILs-MAS (Valdés-Martini et al., 2012) was employed to calculate the topological descriptors named atomic quadratic indices $q_k(x)$ (Marrero-Ponce, 2003; Marrero-Ponce et al., 2003, 2004, 2005a,b, 2011; Montero-Torres et al., 2006). These molecular

TABLE 3.1 Cutoff Values Used to Annotate the Molecules as Positive

Measure of Effect (Units)	Biological Profile	Definition	Cutoff
EC_{50} (nM)	Anti-HIV activity	Concentration required for inhibition of HIV replication by 50% in infected cells	≤800
EC_{90} (nM)	Anti-HIV activity	Concentration required for inhibition of HIV replication by 90% in infected cells	≤1400
AUC (μM.h) ip-LA	ADMET (bioavailability, elimination)	Area under the curve after intraperitoneal administration in laboratory animals	≥18.33
AUC (μM.h) iv-LA	ADMET (bioavailability, elimination)	Area under the curve after intravenous administration in laboratory animals	≥10
AUC (μM.h) po-LA	ADMET (bioavailability, elimination)	Area under the curve after oral administration in laboratory animals	≥12
$t_{1/2}$ (h)ip-LA	ADMET (elimination)	Half-life after intraperitoneal administration in laboratory animals	≥1
$t_{1/2}$ (h)iv-LA	ADMET (elimination)	Half-life after intravenous administration in laboratory animals	≥1.5
$t_{1/2}$ (h)po-LA	ADMET (elimination)	Half-life after oral administration in laboratory animals	≥2.7
F (%)po-LA	ADMET (bioavailability)	Oral bioavailability assessed as the fraction of an oral administered drug that reaches systemic circulation in laboratory animals	≥65
AUC (μM.h) iv-H	ADMET (bioavailability, elimination)	Area under the curve after intravenous administration in humans	≥247.43
AUC (μM.h) po-H	ADMET (bioavailability, elimination)	Area under the curve after oral administration in humans	≥60.34
F (%)po-H	ADMET (bioavailability)	Oral bioavailability assessed as the fraction of an orally administered drug that reaches systemic circulation in humans	≥60
$t_{1/2}$ (h)iv-H	ADMET (elimination)	Half-life after intravenous administration in humans	≥6.9

TABLE 3.1 Cutoff Values Used to Annotate the Molecules as Positive—cont'd

Measure of Effect (Units)	Biological Profile	Definition	Cutoff
$t_{1/2}$ (h)po-H	ADMET (elimination)	Half-life after oral administration in humans	≥4.53
Vdss (L/kg)-H	ADMET (distribution)	Volume of distribution at steady state after intravenous administration in humans	≥0.45
Permeability (nm/s)	ADMET (absorption)	Permeability in cells	≥130
PPB (%)-H	ADMET (distribution)	Plasma protein binding in humans	≤50
LD_{50} (μmol/kg)im	ADMET (toxicity)	Lethal dose at 50% after intramuscular administration in laboratory animals	≥580
LD_{50} (μmol/kg)ip	ADMET (toxicity)	Lethal dose at 50% after intraperitoneal administration in laboratory animals	≥700
LD_{50} (μmol/kg)iv	ADMET (toxicity)	Lethal dose at 50% after intravenous administration in laboratory animals	≥400
LD_{50} (μmol/kg)po	ADMET (toxicity)	Lethal dose at 50% after oral administration in laboratory animals	≥2402.74
LD_{50} (μmol/kg)sc	ADMET (toxicity)	Lethal dose at 50% after subcutaneous administration in laboratory animals	≥639.55
TD_{50} (μmol/kg)ip	ADMET (toxicity)	Dose causing toxic effects after intraperitoneal administration in 50% of the laboratory animals used in the assays	≥390.4
TD_{50} (μmol/kg)po	ADMET (toxicity)	Dose causing toxic effects after oral administration in 50% of the laboratory animals used in the assays	≥911.3
IC_{50} (nM)cyp	ADMET (metabolism)	Concentration that inhibits 50% of the enzymatic activity of a cytochrome	≥5000
Ki (nM)trp	ADMET (metabolism)	Inhibition constant of a chemical against a transporter protein	≥1610
IC_{50} (nM)trf	ADMET (metabolism)	Concentration that inhibits 50% of the activity of a transferase protein	≥6110

ADMET, absorption, distribution, metabolism, elimination, and toxicity; *HIV*, human immunodeficiency virus.

descriptors are based on the application of discrete mathematics and linear algebra theory to chemistry, and they have been widely used in different fields of research in drug discovery (Casanola-Martin et al., 2007, 2010, 2008; Castillo-Garit et al., 2008; Ibarra-Velarde et al., 2008; Marrero-Ponce, 2003, 2004; Marrero-Ponce et al., 2007, 2006, 2011). Quadratic indices can be calculated according to the following expression:

$$q_k(\bar{x}) = [X^T][\mathrm{M}]^k[X] = \sum_{i=1}^{n}\sum_{j=1}^{n} {}^k m_{ij} \cdot x_i \cdot x_j \tag{3.1}$$

In Eq. (3.1), $\bar{x}$ is a molecular vector and $[X^T]$ is the transpose of $[X]$, which is a column vector ($n \times 1$ matrix) with components $x_1, \ldots, x_n$. In addition, the term ${}^k m_{ij}$ represents the elements of the *kth* power of the adjacency matrix **M** of the molecular pseudograph. The components x of the column vector $[X]$ are used as atomic physicochemical properties (*PP*). Therefore, the general symbols of the quadratic indices can be written as $q_k(PP)$. In this study, we generated a new set of molecular descriptors:

$$Mq_k(PP) = \frac{q_k(PP)}{nAT} \tag{3.2}$$

In Eq. (3.2), $Mq_k(PP)$ represents the modified quadratic index, while *nAT* is the number of atoms of the molecule (hydrogen atoms included), and it was calculated using the program DRAGON (Talete-srl, 2015). Notice that Eq. (3.2) is applied to normalize the values of the quadratic indices, eliminating any possible excessive influence of the molecular size.

3.2.2. Creation of the mtk-QSBER Model

A close inspection of Eq. (3.2) reveals that the $Mq_k(PP)$ indices only consider the chemical structures of the molecules and are not able to account for the different aspects (m_e, b_t, a_i, t_m, and p_c) of the multiple experimental conditions under which a molecule can be assayed. The Box–Jenkins approach, which involves the calculation of moving averages offers a simple solution to this issue (Hill and Lewicki, 2006). Currently, Box–Jenkins operators have been used in several works focused on drug discovery (Romero-Duran et al., 2016; Speck-Planche et al., 2016). In the first step regarding the calculation of the Box–Jenkins moving averages, the following mathematical formalism is used:

$$avg\_Mq_k(PP)c_j = \frac{1}{n(c_j)} \sum_{i=1}^{n(c_j)} Mq_k(PP)_i \tag{3.3}$$

In Eq. (3.3), $n(c_j)$ is the number of molecules assayed by considering the same element of the experimental condition (ontology) c_j, being those molecules annotated as positive. For instance, in the case of the element m_e, $n(c_j)$ will be the number of molecules assigned as positive cases $[BE_i(c_j) = 1]$ that have been assayed by considering the same measure of biological effect. The calculations and deductions derived from Eq. (3.3) were applied to the elements m_e, b_t, a_i, and t_m. In this same equation, $Mq_k(PP)_i$ is the modified quadratic index of the *i*th molecule, while $avg\_Mq_k(PP)c_j$ is the arithmetic mean of the $Mq_k(PP)_i$ indices. After calculating the $avg\_Mq_k(PP)c_j$ values, a subsequent expression can be employed to generate a new type of quadratic index for each molecule:

$$DMq_k(PP)_i c_j = p_c \cdot \left[Mq_k(PP)_i - avg\_Mq_k(PP)c_j\right] \tag{3.4}$$

In Eq. (3.4), the deviation term $DMq_k(PP)_ic_j$ is an adaptation of the Box−Jenkins operators, and it considers both the chemical structure and a specific element of the experimental condition c_j under which a molecule was tested. The probabilistic term p_c has been explained earlier. Only the $DMq_k(PP)_ic_j$ indices (128 in total) were taken into account during the generation of the mtk-QSBER model.

The data set composed of the 29,682 cases was randomly divided into two series: training and prediction (test) sets. The training set was used to seek for the best model, and it was formed by 22,352 cases, 8941 annotated as positive and 13,411 as negative. On the other hand, the prediction set was used to demonstrate the predictive power of the model. This set was formed by 7330 cases, 2902 considered as positive and 4428 negative. Linear discriminant analysis (LDA) based on a forward stepwise procedure was used as the data analysis method for the creation of the mtk-QSBER model (Hill and Lewicki, 2006). The data analysis was performed by the software STATISTICA (Statsoft-Team, 2001). Correlations between variables were also taken into account. Thus, the cutoff interval $-0.75 < r < 0.75$ was applied as a criterion for lack of redundancy, r being the Pearson's correlation coefficient (Pearson, 1895). The general discriminant equation of the mtk-QSBER model can be expressed in the following form:

$$BE_i(c_j) = a_0 + \sum_{m=1}^{z} b_m \cdot \left[DMq_k(PP)_i c_j\right]_m \tag{3.5}$$

In Eq. (3.5), a_0 is the constant term, while b_m refers to the coefficients of the molecular descriptors. It is necessary to emphasize that during the generation of the model, STATISTICA takes the initial categorical values of $BE_i(c_j)$ and transforms them into scores (continuous values) of biological effects. Then, after the application of default procedures used by this program with the aim of finding the best discriminant function, each predicted continuous score is converted to its corresponding categorical value [$Pred\text{-}BE_i(c_j)$].

The quality of the model was assessed through the analysis of several statistical indices such as Wilks's lambda (λ), chi-square (χ^2), *p*-value, which are calculated only for the training set. Additionally, other indices such as sensitivity (percentages of correct classification for positive cases), specificity (percentages of correct classification for negative cases), accuracy (overall percentage of correct classification), Matthews's correlation coefficient (MCC),

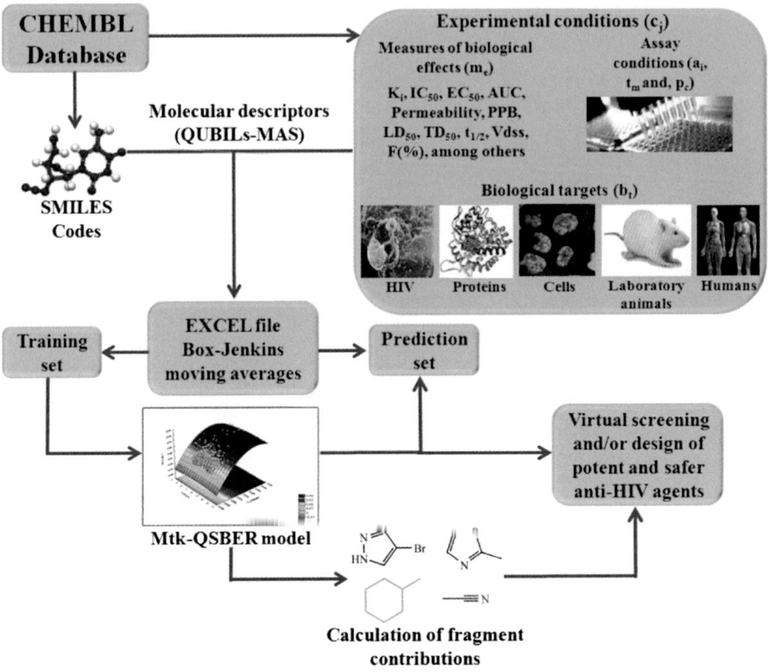

FIGURE 3.1 Different stages involved in the development of the multitasking model for quantitative structure–biological effect relationships.

and the areas under the receiver–operating characteristic (ROC) curves were calculated (Speck-Planche et al., 2016). The last five statistical indices served to demonstrate the internal performance (training set) and the predictive power (prediction set) of the mtk-QSBER model. The general steps involved in the creation of the mtk-QSBER model are illustrated in Fig. 3.1.

3.3. RESULTS AND DISCUSSION

3.3.1. mtk-QSBER Model

For the selection of the most appropriate mtk-QSBER model, we applied the principle of parsimony, which means that only the model with the best combination of high statistical quality and low number of descriptors was chosen. The best model found by us contains seven descriptors:

$$\begin{aligned} BE_i(c_j) = {} & 8.15DMq_1(HYD)m_e - 0.42DMq_4(HYD)m_e - 1.57DMq_0(AR)b_t \\ & -2.27\cdot 10^{-3}DMq_5(PSA)a_i + 16.66DMq_0(HYD)t_m \\ & + 2.08DMq_2(AW)t_m + 0.02DMq_2(PSA)t_m + 1.29 \end{aligned} \tag{3.6}$$

$N = 22,352 \quad \lambda = 0.47 \quad \chi^2 = 17,096.75 \quad p - \text{value} < 10^{-16}$

By simple inspection of Eq. (3.6), one can see that the relatively small value of λ, large χ^2, and the very low p-value indicate the great internal quality of the model. Additionally, the mtk-QSBER model could correctly classify 8627 out of 8941 positive cases (sensitivity = 96.49%) in the training set. In the same set, 12,905 out of 13,411 negative cases (specificity = 96.23%) were rightly classified. The model exhibited an accuracy of 96.33%. At the same time, 2795 of 2902 positive (sensitivity = 96.31%) and 4256 of 4428 negative (specificity = 96.12%) cases were correctly classified in the prediction (test) set, which yielded an accuracy of 96.19%. All the chemical and biological data, as well as the results of the classification for each molecule are represented in the supplementary material (**SM1.xlsx**). In addition to the global measures of correct classification such as sensitivity and specificity, we also calculated local metrics named percentages of correctly classified compounds depending on each specific element of the experimental condition/ontology c_j [$\%CCC(c_j)$]. Thus, for the case of the measures of biological effects (element m_e), its corresponding statistic [$\%CCC(m_e)$] was in the interval 77.78−100%. Specific details of all the metrics of the type $\%CCC(c_j)$ can be found in the supplementary material (**SM2.xlsx** file). It should be pointed out that all the supplementary material files can be accessed via the companion site associated with this book.

Another interesting statistical index was the MCC, which had a value of 0.92 for both training and prediction sets. This means that there is a very strong correlation between the observed and predicted values of the categorical variable of biological effect [$BE_i(c_j)$]. The areas under the ROC curves were also used to demonstrate the performance of the mtk-QSBER model (Fig. 3.2), and they exhibited values of 0.996 and 0.995 for the training and prediction sets, respectively. These values indicate that the model does not behave as a random classifier, for which the value of area under the ROC curve is equal to 0.5. Altogether, the analysis of the diverse statistical indices confirms the high quality and predictive power of the mtk-QSBER model.

3.3.2. Molecular Descriptors and Their Meanings From a Physicochemical Point of View

Most of the chemoinformatic models reported in the literature are mainly used as tools to perform virtual screening of large libraries of molecules. Thus, no attention is paid to the fact that important physicochemical information can be gathered from these models. A correct interpretation of the molecular descriptors in a model can guide the analyst in the right direction regarding the physicochemical and/or structural characteristics that a molecule should possess to improve a defined biological effect (increment of activity, diminution of toxicity, etc.). To accomplish the task of interpreting the molecular descriptors that entered in the mtk-QSBER model, we will rely on their relative influences, which can be estimated through the calculation of the

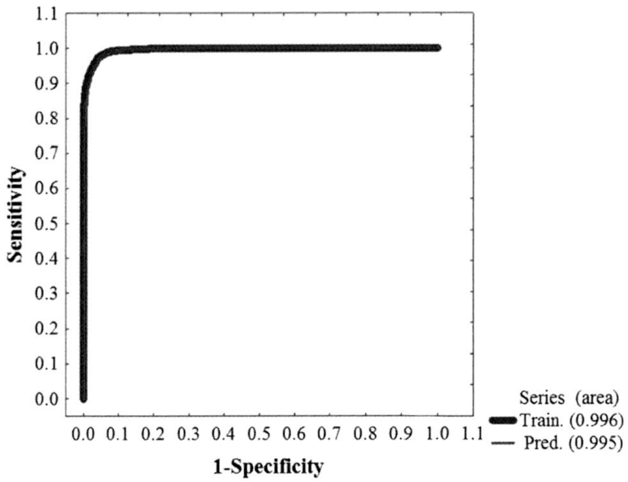

FIGURE 3.2 Areas under the receiver–operating characteristic curves.

absolute values of the standardized coefficients (Fig. 3.3). Here, we will obtain information regarding how the different molecular descriptors should be varied to enhance the anti-HIV activity and increase the safety profiles (ADMET) of any molecule.

The molecular descriptors of type $DMq_k(PP)_ic_j$ are weighted by the squares of the different *PPs*, and the letter *k* represents their order, i.e., the topological distance between any two atoms. As commented before, these descriptors account for the chemical structures of the molecules, and specific elements of the experimental conditions under which the molecules were tested. Another important detail is that $DMq_k(PP)_ic_j$ indices also consider the bond multiplicity. By inspecting Eq. (3.6) and Fig. 3.3, one can see that $\boldsymbol{DMq_5(PSA)a_i}$ is the most influential descriptor in the mtk-QSBER model, expressing the diminution of the polar surface area in regions where atoms are placed at topological distance equal to 5. This molecular descriptor depends on the molecular structure and the information related to a defined assay under which a molecule was tested. Notice that $\boldsymbol{DMq_5(PSA)a_i}$ is constrained by $\boldsymbol{DMq_2(PSA)t_m}$ (the fourth most important variable), which characterizes the increment of the polar surface area in those regions where the atoms are placed at topological distance equal to 2. Here, $\boldsymbol{DMq_2(PSA)t_m}$ depends on the chemical structure and the target mapping.

Molecular descriptors with steric nature also have a great importance. Thus, $\boldsymbol{DMq_2(AW)t_m}$ is a descriptor that embodies the chemical structure and the target mapping, indicating the augmentation of the atomic weight in regions where the atoms are placed at topological distance equal to 2. The

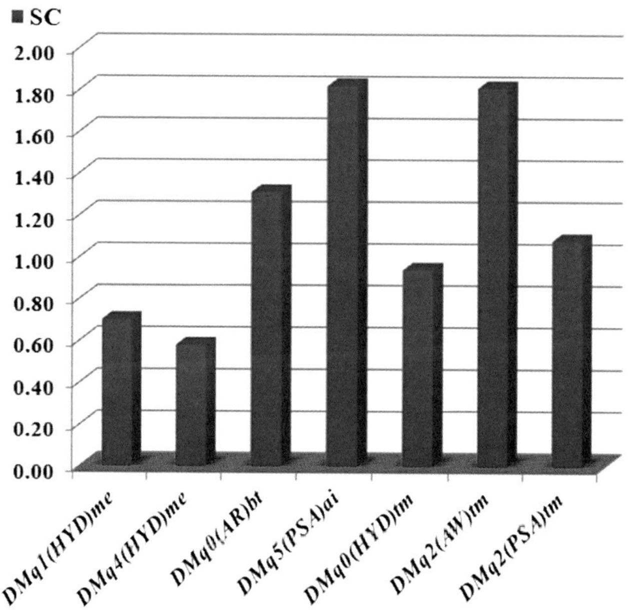

FIGURE 3.3 Standardized coefficients used as measures of the relative significances of the molecular descriptors in the multitasking model for quantitative structure–biological effect relationships.

descriptor $\boldsymbol{DMq_2(AW)t_m}$ is the second most significant in the mtk-QSBER model. On the other hand, $\boldsymbol{DMq_0(AR)b_t}$ has a great influence too (the third most significant variable), and it describes the diminution of the global polarizability without considering the chemical environments of the atoms in the molecules, depending on the chemical structures and the biological targets against which the molecules were assayed.

As commented earlier, $DMq_k(PP)_ic_j$ are quadratic functions of the diverse *PPs*, and bearing in mind that hydrophobicity can take both positive and negative values, this property should be interpreted only in terms of the absolute values. The remaining three descriptors explain variations in the hydrophobicity, which is one of the most important properties in drug discovery. In this sense, $\boldsymbol{DMq_1(HYD)m_e}$ refers to the increase in the global hydrophobicity by considering all the possible pairs of adjacent atoms in a molecule. In the mtk-QSBER model, $\boldsymbol{DMq_1(HYD)m_e}$ is the sixth most influential descriptor, depending on the chemical structure and the measures of biological effects. In convergence with this previous descriptor, $\boldsymbol{DMq_0(HYD)t_m}$ (the fifth most significant) also represents the increment of the global hydrophobicity but without considering the chemical environments of the atoms in the molecules. This descriptor accounts for the chemical structure and the target mapping. Finally, $\boldsymbol{DMq_4(HYD)m_e}$ constrains the other two hydrophobicity-based descriptors because it expresses the decrease of the hydrophobicity in regions where atoms are placed at topological distance equal to 4. It should be noticed that $\boldsymbol{DMq_4(HYD)m_e}$ has the lowest influence in the mtk-QSBER model, and it depends on the structures of the molecules and the different measures of biological effects.

3.3.3. Contribution of Fragments to Multiple Biological Effects

It has been demonstrated that each topological index can be expressed as a linear combination of occurrence numbers of substructures/fragments (Baskin et al., 1995). This envisages the fact that any regression or classification model can be represented as a linear equation involving fragment descriptors (Speck-Planche and Cordeiro, 2013, 2014), leading to the estimation and analysis of the quantitative contributions of fragments to multiple biological effects. Several researchers have emphasized that the calculation of quantitative contributions can serve as a guide for medicinal and pharmaceutical chemists, enabling the detection of 2D pharmacophores, toxicophores, metabolophores, etc. (Estrada et al., 2000; Speck-Planche and Cordeiro, 2013, 2014).

Our mtk-QSBER model based on LDA has been created from the topological descriptors named quadratic indices, and we have used this model as a tool to determine the relative influence of several molecular fragments on the biological effects by considering all the experimental conditions (combinations of the elements m_e, b_t, a_i, t_m, and p_c) reported in our data set (290 in total). In the present study, 40 different molecular fragments were selected (Fig. 3.4), and to calculate their molecular descriptors, Eqs. (3.1)−(3.5) were applied.

Thereafter, the molecular descriptors of the fragments were substituted in Eq. (3.6), which yielded $290 \times 40 = 11{,}600$ "nonstandardized" scores of biological effects. Finally, the mean and the standard deviation of all these scores were calculated, and a subsequent standardization procedure was used, where each standardized score was calculated in the following way: the mean was subtracted from the nonstandardized score, and the result of this subtraction was divided by the standard deviation. The standardized scores represent the quantitative contributions of the fragments to the different biological effects. A sample of the experimental conditions used for the calculation of quantitative contributions is depicted in Tables 3.2 and 3.3. At the same time, the complete list of values of quantitative contributions of the fragments to the diverse biological effects can be found in the supplementary material (**SM3.xlsx**). This file can be accessed via the companion site associated with this book.

In Fig. 3.4, one can see that the molecular fragments are represented with different colors, which are related to their corresponding contributions

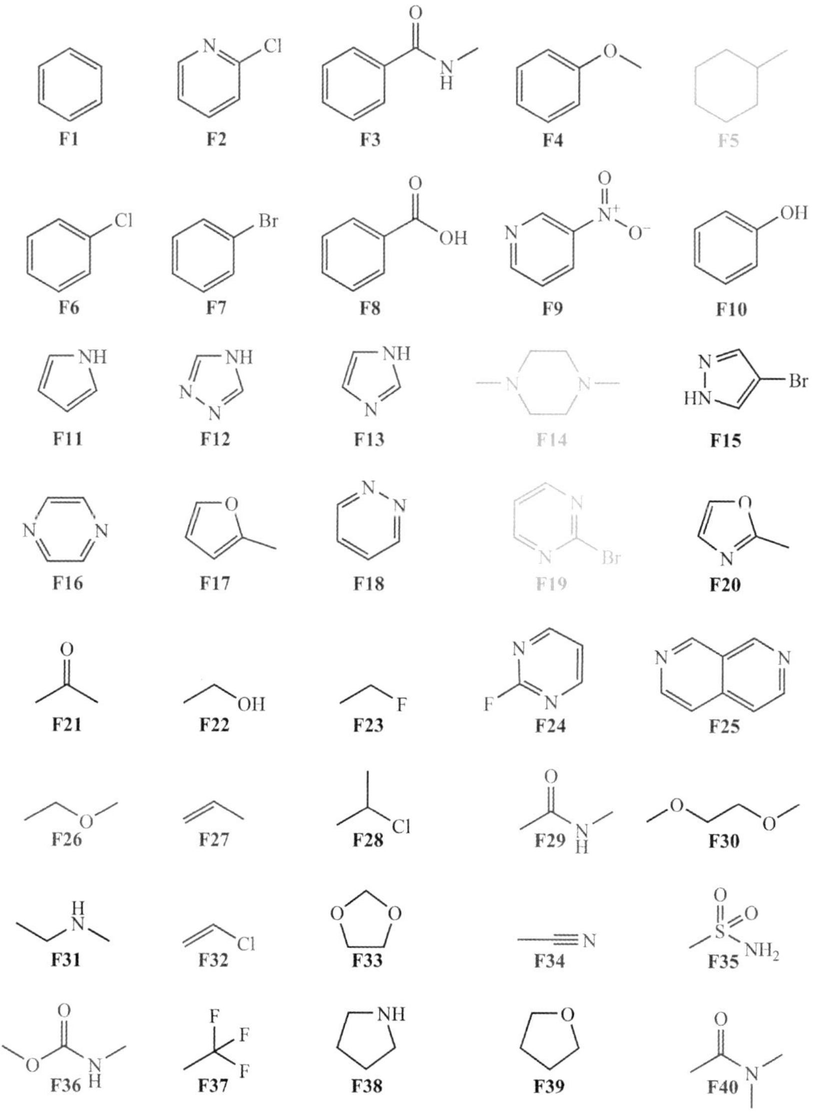

FIGURE 3.4 Fragments selected for the calculation and analysis of the quantitative contributions in multiple experimental conditions.

reported in Table 3.3. Fragments in red color are those that have negative contributions against most of the experimental conditions. Green-colored fragments have positive contributions in a maximum of 70% of the experimental conditions, while fragments with blue color are positive in 80% of the same conditions. Finally, fragments with purple color have positive contributions against 100% of the experimental conditions. Notice that this

TABLE 3.2 Diverse Experimental Conditions Under Which the Fragment Contribution Were Determined

c_j[a,b]	m_e	b_t	a_i	t_m
c_1	EC_{50} (nM)	HIV-1 (IIIB)	Functional	Non-molecular
c_2	Permeability (nm/s)	Caco-2 (Homo sapiens)	ADME	Non-molecular
c_3	F (%)po-H	*Homo sapiens*	ADME	Animal
c_4	F (%)po-LA	*Mus musculus*	ADME	Animal
c_5	AUC (μM.h)po-LA	*M. musculus* (BALB/c)	ADME	Animal
c_6	$t_{1/2}$ (h)po-H	*Homo sapiens*	ADME	Animal
c_7	IC_{50} (nM)cyp	Cytochrome P450 17A1	Binding	Protein
c_8	IC_{50} (nM)cyp	Cytochrome P450 3A4	Binding	Protein
c_9	Ki (nM)trp	Vesicular acetylcholine transporter	Binding	Protein
c_{10}	LD_{50} (μmol/kg)po	*Rattus norvegicus* (Sprague–Dawley)	Toxicity	Animal

ADME, absorption, distribution, metabolism, and elimination; *HIV*, human immunodeficiency virus.
[a] *The symbol c_j is used to represent an ontology/condition as result of the combinations of the elements m_e, b_t, a_i, and t_m.*
[b] *The reliability of all the experimental conditions is assumed to have the label "Expert", for which the associated probabilistic factor p_c has value equal to 1.*

distinction in colors has only been made to compare the different fragments according to their quantitative contributions represented in Table 3.3. Thus, it is mandatory to emphasize that even the green-colored fragments, which seem to be the less favored in the group of active fragments, have positive contributions in at least 259 out of 290 (89.31%) of all the experimental conditions reported in the data set (see **SM3** file). Nevertheless, Table 3.3 provides a guide regarding the fragments that have favorable and detrimental influence in the different biological effects. In this sense, the greater the quantitative contribution of a fragment, the higher is its desirability to be used in the future design of new molecules.

3.3.4. In Silico Design and Screening of Potentially Efficient and Safe Anti-HIV Molecules

In modern drug discovery, current chemoinformatic models are mostly treated as "black boxes" with the only objective of performing virtual screening of large chemical libraries. A big problem and tough challenge for the scientific

TABLE 3.3 Quantitative Contributions of the Fragments to the Different Biological Effects

ID[a]	c_1	c_2	c_3	c_4	c_5	c_6	c_7	c_8	c_9	c_{10}
F1	−1.70	−1.95	−1.53	−1.58	−1.65	−1.51	−2.40	−1.76	−2.43	−1.59
F2	−1.13	−1.38	−0.95	−1.01	−1.08	−0.94	−1.83	−1.19	−1.86	−1.02
F3	−0.19	−0.44	−0.01	−0.07	−0.14	0.00	−0.89	−0.25	−0.92	−0.08
F4	−0.92	−1.17	−0.75	−0.81	−0.87	−0.73	−1.62	−0.98	−1.65	−0.81
F5	0.18	−0.07	0.35	0.29	0.23	0.36	−0.52	0.11	−0.55	0.28
F6	−1.41	−1.66	−1.24	−1.30	−1.36	−1.23	−2.11	−1.48	−2.14	−1.31
F7	−0.57	−0.82	−0.40	−0.46	−0.52	−0.39	−1.27	−0.64	−1.30	−0.47
F8	−0.40	−0.65	−0.23	−0.29	−0.35	−0.21	−1.10	−0.46	−1.13	−0.29
F9	−1.01	−1.26	−0.84	−0.90	−0.96	−0.83	−1.71	−1.08	−1.74	−0.91
F10	−1.47	−1.71	−1.29	−1.35	−1.42	−1.28	−2.17	−1.53	−2.20	−1.36
F11	−0.96	−1.21	−0.79	−0.85	−0.91	−0.78	−1.66	−1.03	−1.69	−0.86
F12	−1.12	−1.37	−0.94	−1.00	−1.07	−0.93	−1.82	−1.18	−1.85	−1.01
F13	−0.11	−0.36	0.06	0.00	−0.06	0.08	−0.81	−0.17	−0.84	0.00
F14	0.20	−0.05	0.37	0.31	0.25	0.38	−0.50	0.13	−0.53	0.30
F15	0.35	0.10	0.53	0.47	0.40	0.54	−0.35	0.29	−0.38	0.46
F16	−1.11	−1.36	−0.94	−1.00	−1.06	−0.93	−1.81	−1.17	−1.84	−1.00

Continued

TABLE 3.3 Quantitative Contributions of the Fragments to the Different Biological Effects—cont'd

ID[a]	c_1	c_2	c_3	c_4	c_5	c_6	c_7	c_8	c_9	c_{10}
F17	−0.69	−0.94	−0.51	−0.57	−0.64	−0.50	−1.39	−0.75	−1.42	−0.58
F18	−1.59	−1.84	−1.42	−1.48	−1.54	−1.41	−2.29	−1.65	−2.32	−1.48
F19	0.07	−0.18	0.24	0.18	0.12	0.25	−0.63	0.00	−0.66	0.17
F20	0.48	0.23	0.65	0.59	0.53	0.66	−0.22	0.41	−0.25	0.58
F21	0.54	0.29	0.71	0.65	0.59	0.72	−0.16	0.47	−0.19	0.64
F22	0.69	0.44	0.86	0.80	0.74	0.87	−0.01	0.62	−0.04	0.79
F23	0.59	0.34	0.76	0.70	0.64	0.77	−0.11	0.53	−0.14	0.70
F24	−0.72	−0.97	−0.55	−0.61	−0.67	−0.54	−1.42	−0.79	−1.45	−0.62
F25	−0.92	−1.17	−0.75	−0.81	−0.87	−0.74	−1.62	−0.99	−1.65	−0.82
F26	0.74	0.49	0.91	0.85	0.79	0.92	0.04	0.67	0.01	0.84
F27	−0.60	−0.84	−0.42	−0.48	−0.55	−0.41	−1.30	−0.66	−1.32	−0.49
F28	0.27	0.02	0.44	0.38	0.32	0.45	−0.43	0.20	−0.46	0.37
F29	1.11	0.86	1.28	1.22	1.16	1.29	0.41	1.04	0.38	1.21
F30	0.64	0.39	0.81	0.75	0.69	0.82	−0.06	0.57	−0.09	0.74
F31	0.64	0.39	0.81	0.75	0.69	0.82	−0.06	0.57	−0.09	0.74
F32	−0.63	−0.88	−0.46	−0.52	−0.58	−0.45	−1.33	−0.69	−1.36	−0.52

F33	0.48	0.23	0.66	0.60	0.53	0.67	−0.22	0.42	−0.25	0.59
F34	3.26	3.01	3.44	3.38	3.31	3.45	2.56	3.20	2.53	3.37
F35	0.00	−0.24	0.18	0.12	0.05	0.19	−0.70	−0.06	−0.72	0.11
F36	1.63	1.38	1.80	1.74	1.68	1.81	0.93	1.56	0.90	1.73
F37	0.45	0.21	0.63	0.57	0.50	0.64	−0.25	0.39	−0.28	0.56
F38	0.30	0.05	0.47	0.41	0.35	0.49	−0.40	0.24	−0.43	0.41
F39	0.41	0.16	0.58	0.52	0.46	0.59	−0.29	0.34	−0.32	0.51
F40	1.05	0.80	1.23	1.17	1.10	1.24	0.35	0.99	0.32	1.16

[a] *Referred to the same experimental conditions represented in Table 3.2.*

community is to assess the confidence of the predictions. Thus, concepts such as applicability domain have been developed as an alternative to create more reliable in silico tools (Carrio et al., 2014; Gaspar et al., 2013; Sahigara et al., 2012; Toplak et al., 2014). However, regardless of the use of any methodology, data analysis method, and applicability domain approach, chemoinformatic models will always tend to fail to some extent when performing virtual predictions.

There are several important factors associated with this issue. A first factor is the lack of reliability of the experimental results, which emerges from the exponential accumulation of assay outcomes from many research laboratories worldwide. Each laboratory has its own experimental error, and therefore, it is extremely difficult to control the uncertainty of the data when creating a model. A second factor is that there is no universal molecular descriptor. This means that each molecular descriptor can characterize a very small portion of all the diversity and complexity enclosed within the molecules. Therefore, even with the combination of many descriptors having different physicochemical meanings, the model derived from them will be capable of correctly mining a relatively reduced chemical space. A third factor is that in most of the chemoinformatic models reported in the literature, the molecular descriptors are only focused on aspects of the chemical structure, and just a few attempts (when compared with the extensive literature) have been made to introduce more biological information inside the molecular descriptors used to create the chemoinformatic models (Gonzalez-Diaz et al., 2006, 2014; Herrera-Ibata et al., 2014, 2015; Prado-Prado et al., 2007, 2008, 2009; Romero-Duran et al., 2016; Speck-Planche and Cordeiro, 2014, 2015; Speck-Planche and Kleandrova, 2012; Speck-Planche et al., 2012, 2016; Tenorio-Borroto et al., 2014). Consequently, this lack of biological meaning in the molecular descriptors will undermine the reliability of any model.

Bearing in mind all these ideas, in this chapter we use the mtk-QSBER model as a knowledge generator, i.e., as a tool to generate new molecular entities from some of the fragments with positive contributions against most of (or all) the experimental conditions. There are several reasons that make us believe this can be a promising alternative in drug discovery, and in this particular study, toward the design of new, and potentially active and safe anti-HIV agents. From one side, due to its simplicity, molecular fragments are usually easier to characterize by any descriptor than the entire molecules they come from. On the other hand, the generation of the new molecules will be performed by considering the quantitative contributions of the fragments to multiple biological effects, and the physicochemical interpretations of the molecular descriptors in the mtk-QSBER model. In this subsection, we intend to demonstrate that at least from a theoretical point of view, potent and desirable anti-HIV drug candidates can be designed (Fig. 3.5).

We designed six molecules using combinations of different fragments with positive contributions. Such combinations were realized via connecting the

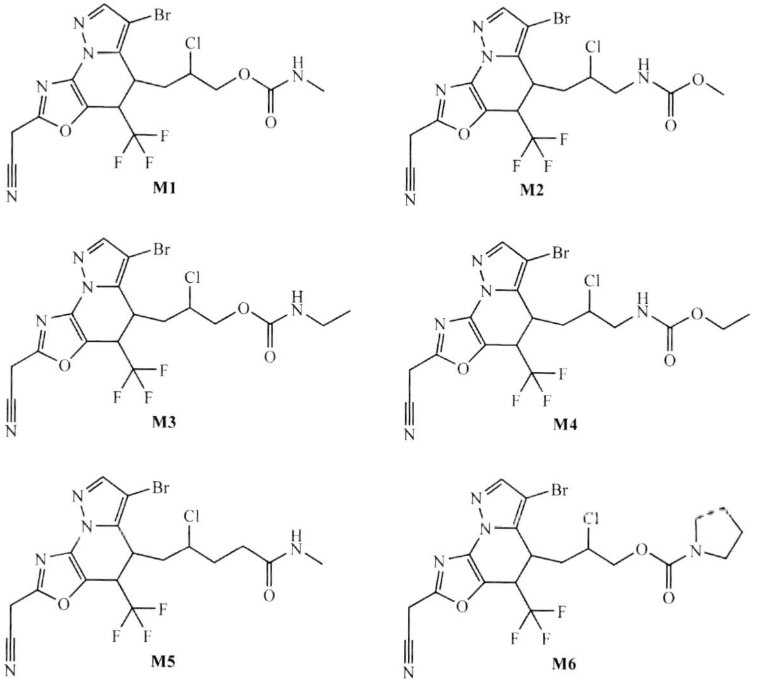

FIGURE 3.5 New molecules designed by the multitasking model for quantitative structure–biological effect relationships.

aforementioned fragments or fusing them. The designed molecules were predicted against 870 diverse experimental conditions [multiplying the 290 original conditions by the three levels of reliability of the assays ($p_c = 0.55$, $p_c = 0.75$, and $p_c = 1$). The results of the virtual predictions can be found in the supplementary material (**SM4.xlsx**), which can be accessed via the companion website associated with this book. In addition, a summary of the predictions involving all the assay conditions reported in Table 3.2 are represented in Table 3.4.

The inspection of Table 3.2, and the analyses of the results of the predictions depicted in Table 3.4 and the **SM4.xlsx** file suggest that the six designed molecules have high anti-HIV activity, good permeability, and oral bioavailability in both *in vitro* and *in vivo* models. These new molecules were also predicted to exhibit large elimination half-lives and low toxicity. In terms of metabolism, the designed molecules were classified as positive, which means that they do not seem to be strong inhibitors of major drug-metabolizing enzymes such as cytochrome P450 3A4. However, the screening performed by the mtk-QSBER model indicates that four of these molecules were predicted as negative against cytochrome P450 17A1 (Tables 3.2 and 3.4), and cytochrome P450 26A1 (**SM4.xlsx** file), and

TABLE 3.4 Summary of the Predictions Performed by the mtk-QSBER Against Six New Molecular Entities

ID[a]	Probabilities (%)[a,b]									
	c_1	c_2	c_3	c_4	c_5	c_6	c_7	c_8	c_9	c_{10}
M1	99.91	99.28	99.98	99.96	99.94	99.98	59.83	99.85	52.22	99.96
M2	99.89	99.09	99.97	99.96	99.93	99.97	52.75	99.81	45.02	99.95
M3	99.88	98.96	99.97	99.95	99.92	99.97	48.92	99.79	41.28	99.95
M4	99.88	98.96	99.97	99.95	99.92	99.97	48.80	99.79	41.17	99.95
M5	99.47	94.83	99.88	99.80	99.66	99.89	13.64	99.06	10.55	99.78
M6	99.63	96.60	99.91	99.86	99.76	99.92	19.90	99.36	15.63	99.85

mtk-QSBER, multitasking model for quantitative structure–biological effect relationships.
[a]*Probabilities of the molecules to be classified as positive.*
[b]*Referred to the same experimental conditions represented in Table 3.2.*

TABLE 3.5 Different Molecular Properties of the Designed Molecules

	Molecular Properties[a]							
ID	Molar Mass	nAT	RBN	nHDon	nHAcc	AMR	PSA	MlogP
M1	496.7	43	6	1	10	99.11	105.97	2.91
M2	496.7	43	6	1	10	99.11	105.97	2.91
M3	510.73	46	7	1	10	103.86	105.97	3.13
M4	510.73	46	7	1	10	103.86	105.97	3.13
M5	494.73	45	6	1	9	102.35	96.74	3.47
M6	536.77	50	6	0	10	111.55	97.18	3.57

nAT, number of atoms (including hydrogen atoms); *RBN*, number of rotatable bonds; *nHDon*, number of hydrogen bond donors; *nHAcc*, number of hydrogen bond acceptors; *AMR*, molar refractivity; *PSA*, polar surface area, *MlogP*, logarithm of the partition coefficient (octanol/water) according to the Moriguchi approach.
[a]*Fluorine atoms are also considered as hydrogen bond acceptors in the calculation of nHAcc.*

therefore, they may strongly interact with these two proteins. At the same time, all the molecules except **M1** were also predicted as negative (high affinity) against the vesicular acetylcholine transporter (Tables 3.2 and 3.4), glycine transporter 1, and monocarboxylate transporter 1 (see results regarding the last two proteins in the **SM4.xlsx** file). Finally, it should be noticed that all the designed molecules were predicted as positive against transferases, which suggests that they may not be strong inhibitors of these enzymes.

In an attempt to gather more information from the designed molecules, we calculated some of the most common properties used in drug discovery. These properties are summarized in Table 3.5 and were calculated using the software DRAGON (Talete-srl, 2015). From the perspective of a medicinal chemist, and according to Lipinski's rule of five (Lipinski et al., 2001), any molecule considered to be orally administered should possess the following features: (1) no more than five hydrogen bond donors (the total number of nitrogen–hydrogen and oxygen–hydrogen bonds); (2) no more than 10 hydrogen bond acceptors (all nitrogen or oxygen atoms); (3) molecular mass less than 500 Da; and (4) logarithm of the octanol–water partition coefficient (logP) not greater than 5 or Moriguchi's logP not greater than 4.15. A second variant (Ghose et al., 1999) of the aforementioned rule proposes logP in the −0.4 to +5.6 range, molar refractivity from 40 to 130, molecular weight from 180 to 500, number of atoms from 20 to 70, and polar surface area no greater than 140 $Å^2$.

According to Table 3.5, in three of the designed molecules there are no violations of the Lipinski's rule of five, while the other three molecules violate only one rule due to their molecular weights, which are higher than 500 Da.

However, we must emphasize that regarding the molecular weight, Lipinski's rule of five has received some criticism (Veber et al., 2002), and it has been suggested that only molecules with 10 or fewer rotatable bonds and polar surface area equal to or less than 140 $Å^2$ are predicted to have good oral bioavailability. Thus, the joint results of the virtual predictions performed by the mtk-QSBER, and the analysis of the molecular properties commonly calculated in drug discovery programs, permit to consider these six designed molecules as candidates to be used in future biological assays intended to test their anti-HIV activities, as well as their *in vitro* and *in vivo* ADMET properties.

3.4. CONCLUSIONS

The search for new anti-HIV drugs remains a major challenge in the scientific world. Due to the accumulation of huge amounts of experimental data, chemoinformatic models should be employed beyond the classical purpose of screening large data sets. More phenomenological information should be introduced inside the different molecular descriptors in order to create more biologically meaningful models. Special attention should be paid to the fact that the chemoinformatic models may be used as tools to perform true rational design of new therapeutic agents. In this sense, the mtk-QSBER model developed in this chapter constitutes an encouraging alternative in antiviral research toward the discovery of potentially promising anti-HIV drug candidates. The great statistical quality and predictive power of the mtk-QSBER model, together with the physicochemical meanings of the molecular descriptors and the application of an innovative fragment-based topological approach, permitted the guided design of six new molecules that were predicted to display potent anti-HIV activities and desirable ADMET properties. The present mtk-QSBER model creates the foundations to speed up the design and screening of highly efficacious anti-HIV drugs.

ACKNOWLEDGMENTS

The authors are grateful to Prof. Yovani Marrero-Ponce for the creation of the software QUBILs-MAS, which is available to the scientific community via the website http://tomocomd.com/. This is a valuable chemoinformatic tool for drug discovery.

REFERENCES

Afantitis, A., Melagraki, G., Sarimveis, H., Koutentis, P.A., Markopoulos, J., Igglessi-Markopoulou, O., 2006. Investigation of substituent effect of 1-(3,3-diphenylpropyl)-piperidinyl phenylacetamides on CCR5 binding affinity using QSAR and virtual screening techniques. J. Comput. Aided Mol. Des. 20, 83–95.

Arkan, E., Shahlaei, M., Pourhossein, A., Fakhri, K., Fassihi, A., 2010. Validated QSAR analysis of some diaryl substituted pyrazoles as CCR2 inhibitors by various linear and nonlinear multivariate chemometrics methods. Eur. J. Med. Chem. 45, 3394–3406.

Bak, A., Polanski, J., 2006. A 4D-QSAR study on anti-HIV HEPT analogues. Bioorg. Med. Chem. 14, 273–279.

Barreiro, G., Kim, J.T., Guimaraes, C.R., Bailey, C.M., Domaoal, R.A., Wang, L., Anderson, K.S., Jorgensen, W.L., 2007. From docking false-positive to active anti-HIV agent. J. Med. Chem. 50, 5324–5329.

Baskin, I.I., Skvortsova, M.I., Stankevich, I.V., Zefirov, N.S., 1995. On the basis of invariants of labeled molecular graphs. J. Chem. Inf. Comput. Sci. 35, 527–531.

Boutton, C.W., De Bondt, H.L., De Jonge, M.R., 2005. Genotype dependent QSAR for HIV-1 protease inhibition. J. Med. Chem. 48, 2115–2120.

Carrio, P., Pinto, M., Ecker, G., Sanz, F., Pastor, M., 2014. Applicability Domain ANalysis (ADAN): a robust method for assessing the reliability of drug property predictions. J. Chem. Inf. Model. 54, 1500–1511.

Casanola-Martin, G.M., Marrero-Ponce, Y., Khan, M.T., Ather, A., Sultan, S., Torrens, F., Rotondo, R., 2007. TOMOCOMD-CARDD descriptors-based virtual screening of tyrosinase inhibitors: evaluation of different classification model combinations using bond-based linear indices. Bioorg. Med. Chem. 15, 1483–1503.

Casanola-Martin, G.M., Marrero-Ponce, Y., Khan, M.T., Khan, S.B., Torrens, F., Perez-Jimenez, F., Rescigno, A., Abad, C., 2010. Bond-based 2D quadratic fingerprints in QSAR studies: virtual and in vitro tyrosinase inhibitory activity elucidation. Chem. Biol. Drug Des. 76, 538–545.

Casanola-Martin, G.M., Marrero-Ponce, Y., Tareq Hassan Khan, M., Torrens, F., Perez-Gimenez, F., Rescigno, A., 2008. Atom- and bond-based 2D TOMOCOMD-CARDD approach and ligand-based virtual screening for the drug discovery of new tyrosinase inhibitors. J. Biomol. Screen. 13, 1014–1024.

Castillo-Garit, J.A., Marrero-Ponce, Y., Torrens, F., Garcia-Domenech, R., 2008. Estimation of ADME properties in drug discovery: predicting Caco-2 cell permeability using atom-based stochastic and non-stochastic linear indices. J. Pharm. Sci. 97, 1946–1976.

ChemAxon, 1998–2016. Standardizer (tool for structure canonicalization and transformation). JChem v16.7.25.0, Budapest, Hungary.

Estrada, E., Uriarte, E., Montero, A., Teijeira, M., Santana, L., De Clercq, E., 2000. A novel approach for the virtual screening and rational design of anticancer compounds. J. Med. Chem. 43, 1975–1985.

Fauci, A.S., 2008. 25 years of HIV. Nature 453, 289–290.

Fauci, A.S., Folkers, G.K., 2012. Toward an AIDS-free generation. JAMA 308, 343–344.

Fujii, N., Oishi, S., Hiramatsu, K., Araki, T., Ueda, S., Tamamura, H., Otaka, A., Kusano, S., Terakubo, S., Nakashima, H., Broach, J.A., Trent, J.O., Wang, Z.X., Peiper, S.C., 2003. Molecular-size reduction of a potent CXCR4-chemokine antagonist using orthogonal combination of conformation- and sequence-based libraries. Angew. Chem. Int. Ed. 42, 3251–3253.

Gaspar, H.A., Marcou, G., Horvath, D., Arault, A., Lozano, S., Vayer, P., Varnek, A., 2013. Generative topographic mapping-based classification models and their applicability domain: application to the Biopharmaceutics Drug Disposition Classification System (BDDCS). J. Chem. Inf. Model. 53, 3318–3325.

Gaulton, A., Bellis, L.J., Bento, A.P., Chambers, J., Davies, M., Hersey, A., Light, Y., McGlinchey, S., Michalovich, D., Al-Lazikani, B., Overington, J.P., 2012. ChEMBL: a large-scale bioactivity database for drug discovery. Nucleic Acids Res. 40, D1100–D1107.

Ghose, A.K., Viswanadhan, V.N., Wendoloski, J.J., 1999. A knowledge-based approach in designing combinatorial or medicinal chemistry libraries for drug discovery. 1. A qualitative and quantitative characterization of known drug databases. J. Comb. Chem. 1, 55–68.

Gonzalez-Diaz, H., Herrera-Ibata, D.M., Duardo-Sanchez, A., Munteanu, C.R., Orbegozo-Medina, R.A., Pazos, A., 2014. ANN multiscale model of anti-HIV drugs activity vs AIDS prevalence in the US at county level based on information indices of molecular graphs and social networks. J. Chem. Inf. Model. 54, 744–755.

Gonzalez-Diaz, H., Prado-Prado, F.J., Santana, L., Uriarte, E., 2006. Unify QSAR approach to antimicrobials. Part 1: predicting antifungal activity against different species. Bioorg. Med. Chem. 14, 5973–5980.

Granich, R., Gupta, S., Hersh, B., Williams, B., Montaner, J., Young, B., Zuniga, J.M., 2015. Trends in AIDS deaths, new infections and ART Coverage in the top 30 countries with the highest AIDS mortality burden; 1990–2013. PLoS One 10, e0131353.

Gupta, P., Roy, N., Garg, P., 2009. Docking-based 3D-QSAR study of HIV-1 integrase inhibitors. Eur. J. Med. Chem. 44, 4276–4287.

Herrera-Ibata, D.M., Pazos, A., Orbegozo-Medina, R.A., Gonzalez-Diaz, H., 2014. Mapping networks of anti-HIV drug cocktails vs. AIDS epidemiology in the US counties. Chemometr. Intell. Lab. Syst. 138, 161–170.

Herrera-Ibata, D.M., Pazos, A., Orbegozo-Medina, R.A., Romero-Duran, F.J., Gonzalez-Diaz, H., 2015. Mapping chemical structure-activity information of HAART-drug cocktails over complex networks of AIDS epidemiology and socioeconomic data of U.S. counties. Biosystems 132–133, 20–34.

Hill, T., Lewicki, P., 2006. Statistics Methods and Applications. A Comprehensive Reference for Science, Industry and Data Mining. StatSoft, Tulsa.

Ibarra-Velarde, F., Vera-Montenegro, Y., Huesca-Guillen, A., Canto-Alarcon, G., Alcala-Canto, Y., Marrero-Ponce, Y., 2008. In silico fasciolicide activity of three experimental compounds in sheep. Ann. N. Y. Acad. Sci. 1149, 183–185.

Lima, V.D., Gill, V.S., Yip, B., Hogg, R.S., Montaner, J.S., Harrigan, P.R., 2008. Increased resilience to the development of drug resistance with modern boosted protease inhibitor-based highly active antiretroviral therapy. J. Infect. Dis. 198, 51–58.

Lipinski, C.A., Lombardo, F., Dominy, B.W., Feeney, P.J., 2001. Experimental and computational approaches to estimate solubility and permeability in drug discovery and development settings. Adv. Drug Deliv. Rev. 46, 3–26.

Lohse, N., Jorgensen, L.B., Kronborg, G., Moller, A., Kvinesdal, B., Sorensen, H.T., Obel, N., Gerstoft, J., 2007. Genotypic drug resistance and long-term mortality in patients with triple-class antiretroviral drug failure. Antivir. Ther. 12, 909–917.

Luther, J., Glesby, M.J., 2007. Dermatologic adverse effects of antiretroviral therapy: recognition and management. Am. J. Clin. Dermatol. 8, 221–233.

Maartens, G., Celum, C., Lewin, S.R., 2014. HIV infection: epidemiology, pathogenesis, treatment, and prevention. Lancet 384, 258–271.

Marrero-Ponce, Y., 2003. Total and local quadratic indices of the molecular pseudograph's atom adjacency matrix: applications to the prediction of physical properties of organic compounds. Molecules 8, 687–726.

Marrero-Ponce, Y., 2004. Linear indices of the "molecular pseudograph's atom adjacency matrix": definition, significance-interpretation, and application to QSAR analysis of flavone derivatives as HIV-1 integrase inhibitors. J. Chem. Inf. Comput. Sci. 44, 2010–2026.

Marrero-Ponce, Y., Cabrera Pérez, M.A., Romero Zaldivar, V., Ofori, E., Montero, L.A., 2003. Total and local quadratic indices of the "molecular pseudograph's atom adjacency matrix". Application to prediction of Caco-2 permeability of drugs. Int. J. Mol. Sci. 4, 512–536.

Marrero-Ponce, Y., Castillo-Garit, J.A., Olazabal, E., Serrano, H.S., Morales, A., Castanedo, N., Ibarra-Velarde, F., Huesca-Guillen, A., Jorge, E., del Valle, A., Torrens, F., Castro, E.A., 2004. TOMOCOMD-CARDD, a novel approach for computer-aided 'rational' drug design: I. Theoretical and experimental assessment of a promising method for computational screening and in silico design of new anthelmintic compounds. J. Comput. Aided Mol. Des. 18, 615–634.

Marrero-Ponce, Y., Iyarreta-Veitia, M., Montero-Torres, A., Romero-Zaldivar, C., Brandt, C.A., Avila, P.E., Kirchgatter, K., Machado, Y., 2005a. Ligand-based virtual screening and in silico design of new antimalarial compounds using nonstochastic and stochastic total and atom-type quadratic maps. J. Chem. Inf. Model. 45, 1082–1100.

Marrero-Ponce, Y., Khan, M.T., Casanola Martin, G.M., Ather, A., Sultankhodzhaev, M.N., Torrens, F., Rotondo, R., 2007. Prediction of tyrosinase inhibition activity using atom-based bilinear indices. ChemMedChem 2, 449–478.

Marrero-Ponce, Y., Medina-Marrero, R., Torrens, F., Martinez, Y., Romero-Zaldivar, V., Castro, E.A., 2005b. Atom, atom-type, and total nonstochastic and stochastic quadratic fingerprints: a promising approach for modeling of antibacterial activity. Bioorg. Med. Chem. 13, 2881–2899.

Marrero-Ponce, Y., Meneses-Marcel, A., Castillo-Garit, J.A., Machado-Tugores, Y., Escario, J.A., Barrio, A.G., Pereira, D.M., Nogal-Ruiz, J.J., Aran, V.J., Martinez-Fernandez, A.R., Torrens, F., Rotondo, R., Ibarra-Velarde, F., Alvarado, Y.J., 2006. Predicting antitrichomonal activity: a computational screening using atom-based bilinear indices and experimental proofs. Bioorg. Med. Chem. 14, 6502–6524.

Marrero-Ponce, Y., Siverio-Mota, D., Galvez-Llompart, M., Recio, M.C., Giner, R.M., Garcia-Domenech, R., Torrens, F., Aran, V.J., Cordero-Maldonado, M.L., Esguera, C.V., de Witte, P.A., Crawford, A.D., 2011. Discovery of novel anti-inflammatory drug-like compounds by aligning in silico and in vivo screening: the nitroindazolinone chemotype. Eur. J. Med. Chem. 46, 5736–5753.

Mok, N.Y., Brenk, R., 2011. Mining the ChEMBL database: an efficient chemoinformatics workflow for assembling an ion channel-focused screening library. J. Chem. Inf. Model. 51, 2449–2454.

Montero-Torres, A., Garcia-Sanchez, R.N., Marrero-Ponce, Y., Machado-Tugores, Y., Nogal-Ruiz, J.J., Martinez-Fernandez, A.R., Aran, V.J., Ochoa, C., Meneses-Marcel, A., Torrens, F., 2006. Non-stochastic quadratic fingerprints and LDA-based QSAR models in hit and lead generation through virtual screening: theoretical and experimental assessment of a promising method for the discovery of new antimalarial compounds. Eur. J. Med. Chem. 41, 483–493.

Montessori, V., Press, N., Harris, M., Akagi, L., Montaner, J.S., 2004. Adverse effects of antiretroviral therapy for HIV infection. CMAJ 170, 229–238.

Napravnik, S., Keys, J.R., Quinlivan, E.B., Wohl, D.A., Mikeal, O.V., Eron Jr., J.J., 2007. Triple-class antiretroviral drug resistance: risk and predictors among HIV-1-infected patients. AIDS 21, 825–834.

Niedbala, H., Polanski, J., Gieleciak, R., Musiol, R., Tabak, D., Podeszwa, B., Bak, A., Palka, A., Mouscadet, J.F., Gasteiger, J., Le Bret, M., 2006. Comparative molecular surface analysis (CoMSA) for virtual combinatorial library screening of styrylquinoline HIV-1 blocking agents. Comb. Chem. High Throughput Screen. 9, 753–770.

Nolan, D., Reiss, P., Mallal, S., 2005. Adverse effects of antiretroviral therapy for HIV infection: a review of selected topics. Expert Opin. Drug Saf. 4, 201–218.

Overington, J., 2009. ChEMBL. An interview with John Overington, team leader, chemogenomics at the European Bioinformatics Institute Outstation of the European molecular Biology laboratory (EMBL-EBI). Interview by Wendy A. Warr. J. Comput. Aided Mol. Des. 23, 195–198.

Pearson, K., 1895. Notes on regression and inheritance in the case of two parents. Proc. R. Soc. Lond. 58, 240–242.

Perez-Nueno, V.I., Ritchie, D.W., Rabal, O., Pascual, R., Borrell, J.I., Teixido, J., 2008. Comparison of ligand-based and receptor-based virtual screening of HIV entry inhibitors for the CXCR4 and CCR5 receptors using 3D ligand shape matching and ligand-receptor docking. J. Chem. Inf. Model. 48, 509–533.

Prado-Prado, F.J., Gonzalez-Diaz, H., de la Vega, O.M., Ubeira, F.M., Chou, K.C., 2008. Unified QSAR approach to antimicrobials. Part 3: first multi-tasking QSAR model for input-coded prediction, structural back-projection, and complex networks clustering of antiprotozoal compounds. Bioorg. Med. Chem. 16, 5871–5880.

Prado-Prado, F.J., Gonzalez-Diaz, H., Santana, L., Uriarte, E., 2007. Unified QSAR approach to antimicrobials. Part 2: predicting activity against more than 90 different species in order to halt antibacterial resistance. Bioorg. Med. Chem. 15, 897–902.

Prado-Prado, F.J., Martinez de la Vega, O., Uriarte, E., Ubeira, F.M., Chou, K.C., Gonzalez-Diaz, H., 2009. Unified QSAR approach to antimicrobials. 4. Multi-target QSAR modeling and comparative multi-distance study of the giant components of antiviral drug-drug complex networks. Bioorg. Med. Chem. 17, 569–575.

Romero-Duran, F.J., Alonso, N., Yanez, M., Caamano, O., Garcia-Mera, X., Gonzalez-Diaz, H., 2016. Brain-inspired chemoinformatics of drug-target brain interactome, synthesis, and assay of TVP1022 derivatives. Neuropharmacology 103, 270–278.

Sahigara, F., Mansouri, K., Ballabio, D., Mauri, A., Consonni, V., Todeschini, R., 2012. Comparison of different approaches to define the applicability domain of QSAR models. Molecules 17, 4791–4810.

Speck-Planche, A., Cordeiro, M.N.D.S., 2013. Simultaneous modeling of antimycobacterial activities and ADMET profiles: a chemoinformatic approach to medicinal chemistry. Curr. Top. Med. Chem. 13, 1656–1665.

Speck-Planche, A., Cordeiro, M.N.D.S., 2014. Chemoinformatics for medicinal chemistry: in silico model to enable the discovery of potent and safer anti-cocci agents. Future Med. Chem. 6, 2013–2028.

Speck-Planche, A., Cordeiro, M.N.D.S., 2015. Enabling virtual screening of potent and safer antimicrobial agents against noma: mtk-QSBER model for simultaneous prediction of antibacterial activities and ADMET properties. Mini Rev. Med. Chem. 15, 194–202.

Speck-Planche, A., Kleandrova, V.V., 2012. In silico design of multi-target inhibitors for C-C chemokine receptors using substructural descriptors. Mol. Divers. 16, 183–191.

Speck-Planche, A., Kleandrova, V.V., Luan, F., Cordeiro, M.N.D.S., 2012. A ligand-based approach for the in silico discovery of multi-target inhibitors for proteins associated with HIV infection. Mol. Biosyst. 8, 2188–2196.

Speck-Planche, A., Kleandrova, V.V., Ruso, J.M., Cordeiro, M.N.D.S., 2016. First multitarget chemo-bioinformatic model to enable the discovery of antibacterial peptides against multiple gram-positive pathogens. J. Chem. Inf. Model. 56, 588–598.

Statsoft-Team, 2001. STATISTICA. Data analysis software system, v6.0, Tulsa.

Talete-srl, 2015. DRAGON (Software for Molecular Descriptor Calculation), v6.0, Milano.

Tenorio-Borroto, E., Penuelas-Rivas, C.G., Vasquez-Chagoyan, J.C., Castanedo, N., Prado-Prado, F.J., Garcia-Mera, X., Gonzalez-Diaz, H., 2014. Model for high-throughput screening of drug immunotoxicity – Study of the anti-microbial G1 over peritoneal macrophages using flow cytometry. Eur. J. Med. Chem. 72, 206–220.

Toplak, M., Mocnik, R., Polajnar, M., Bosnic, Z., Carlsson, L., Hasselgren, C., Demsar, J., Boyer, S., Zupan, B., Stalring, J., 2014. Assessment of machine learning reliability methods for quantifying the applicability domain of QSAR regression models. J. Chem. Inf. Model. 54, 431–441.

Torres, R.A., Lewis, W., 2014. Aging and HIV/AIDS: pathogenetic role of therapeutic side effects. Lab. Invest. 94, 120–128.

Tozzi, V., Zaccarelli, M., Bonfigli, S., Lorenzini, P., Liuzzi, G., Trotta, M.P., Forbici, F., Gori, C., Bertoli, A., Bellagamba, R., Narciso, P., Perno, C.F., Antinori, A., 2006. Drug-class-wide resistance to antiretrovirals in HIV-infected patients failing therapy: prevalence, risk factors and virological outcome. Antivir. Ther. 11, 553–560.

Valdés-Martini, J.R., García-Jacas, C.R., Marrero-Ponce, Y., Silveira Vaz 'd Almeida, Y., Morell, C., 2012. QUBILs-MAS: Free Software for Molecular Descriptors Calculator from Quadratic, Bilinear and Linear Maps Based on Graph-theoretic Electronic-density Matrices and Atomic Weightings, v1.0, Villa Clara.

Veber, D.F., Johnson, S.R., Cheng, H.Y., Smith, B.R., Ward, K.W., Kopple, K.D., 2002. Molecular properties that influence the oral bioavailability of drug candidates. J. Med. Chem. 45, 2615–2623.

Vedani, A., Dobler, M., Dollinger, H., Hasselbach, K.M., Birke, F., Lill, M.A., 2005. Novel ligands for the chemokine receptor-3 (CCR3): a receptor-modeling study based on 5D-QSAR. J. Med. Chem. 48, 1515–1527.

Waller, C.L., Oprea, T.I., Giolitti, A., Marshall, G.R., 1993. Three-dimensional QSAR of human immunodeficiency virus (I) protease inhibitors. 1. A CoMFA study employing experimentally-determined alignment rules. J. Med. Chem. 36, 4152–4160.

Chapter 4

Alkaloids From the Family Menispermaceae: A New Source of Compounds Selective for β-Adrenergic Receptors

M.F. Alves, M.T. Scotti, L. Scotti, S. Golzio dos Santos, M. de Fátima Formiga Melo Diniz
Federal University of Paraíba, João Pessoa, Paraíba, Brazil

4.1. INTRODUCTION

4.1.1. β-Adrenergic Receptors

The effects of most drugs are attributed to their interactions with macromolecular components of the organism, called pharmacological receptors. These interactions produce biochemical and physiological changes that characterize the effect of the drug (Billingsley, 2008; Wehr and Rossner, 2016).

Pharmacological beta-adrenergic receptors (β-adrenergic receptors) have different tissue distributions and several subtypes. β_1-Adrenergic receptors are found in the heart muscle (Von Homeyer and Schwinn, 2011 and in brain structures such as the cortex, hippoampus, and amygdala (Fitzgerald et al., 2016). β_2-Adrenergic receptors can be found in bronchial tissue (Pelaia et al., 2015), skeletal muscle (Hagg et al., 2016), and cardiomyocytes (Wu et al., 2016). β_3-Adrenergic receptors are mainly found in fatty tissues (Von Homeyer and Schwinn, 2011). The differences in pharmacological effects between subtypes of receptors are exploited therapeutically with the development and utilization of drugs selective for each receptor.

Many drug agonists or antagonists of these receptors play an intrinsic role in the physiology of the organs innervated by the sympathetic nervous system; therefore, they are important in clinical medicine, especially in treating asthmatic (Schofield, 2014) and cardiovascular diseases (Voigt et al., 2014). Pharmacotherapy is often impaired because of the low selectivity of drugs,

Multi-Scale Approaches in Drug Discovery. http://dx.doi.org/10.1016/B978-0-08-101129-4.00004-7

development of side effects, and appearance of drug resistance (Velema et al., 2014). As an example, the higher the selectivity of drugs to β_2-adrenergic agonists, the lower will be the side effects, such as tachycardia and palpitations, by virtue of the nonaction of these drugs on α- and β_1-adrenergic receptors (Schofield, 2014).

Nonselective drugs and selective antagonists of the β_1-adrenergic receptors, such as atenolol (Hoffmann et al., 2004) and carvedilol (Peitzman et al., 2015), are routinely used for the treatment of cardiovascular diseases. For the treatment of asthma, a few selective β_2-adrenergic agonist drugs are used, such as albuterol, levalbuterol, and formoterol (Schofield, 2014).

4.1.2. Family Menispermaceae

The family Menispermaceae is among the 10 major plant families found in most tropical forests, comprising a total of 70 genera (Wefferling et al., 2013). These are characterized by having many secondary metabolites, mainly a wide range of alkaloids, and consequently, several different pharmacological effects that have been assigned (Semwal et al., 2014). The milonina alkaloid could present hypotensive and vasorelaxant effects, which are mediated in part by the release of nitric oxide (NO) in the endothelium (Cavalcante et al., 2010). The alkaloids warifteine and metilwarifteine are considered secondary metabolites with potential antiasthmatic activity inhibiting the release of histamine and reducing the number of $CD3^+$ T cells and eosinophilic cells (Bezerra-Santos et al., 2012; Vieira et al., 2012).

In the search for new selective therapeutic agents that have selective antagonistic effects for selective β_1-adrenergic receptors and agonists for β_2-adrenergic receptors, this study evaluated an in-house data set of alkaloids from the family Menispermaceae, using computer-aided drug design (CADD) methodologies and based on the concentration of an inhibitor where the response (or binding) is reduced by half (IC_{50}) and the concentration of a drug that gives half-maximal response (EC_{50}). For studies with antagonist compounds, the IC_{50} is the most common measure to analyze the dose−response curve, whereas for testing with agonist compounds, the EC_{50} is the most common measure. The EC_{50} and IC_{50} values of the databases were used with selected structures against β_1- and β_2-adrenergic receptors.

4.2. METHODS

4.2.1. Data Set

From the ChEMBL database (https://www.ebi.ac.uk/chembl/), we selected compounds with EC_{50} β_1-adrenergic receptor, EC_{50} β_2-adrenergic receptor, IC_{50} β_1-adrenergic receptor, and IC_{50} β_2-adrenergic receptor values, a set of 342, 99, 373, and 328 structures, respectively. The compounds were classified using values of $-\log EC_{50}$ (mol/L) $= pEC_{50}$ and $-\log IC_{50}$ (mol/L) $= pIC_{50}$. Thus, 169 compounds were assigned as actives ($pEC_{50} \geq 6.5$) and 173 inactives

($pEC_{50} < 6.5$) for EC_{50} β_1-adrenergic receptor, 67 actives ($pEC_{50} \geq 6.5$) and 32 inactives ($pEC_{50} < 6.5$) for EC_{50} β_2-adrenergic receptor, 178 actives ($pIC_{50} \geq 5.7$) and 195 inactives ($pIC_{50} \leq 4$) for IC_{50} β_1-adrenergic receptor, and 190 actives ($pIC_{50} \geq 5.7$) and 138 inactives ($pIC_{50} \leq 4$) for IC_{50} β_2-adrenergic receptor. In this case, IC_{50} represents the concentration required for 50% inhibition of receptor, and it is a measure of antagonistic activity. On the other hand, EC_{50} represents the concentration required for the 50% activation of receptor, and it is a measure of agonistic activity. Our in-house data set consists of secondary metabolites, specifically 786 alkaloids from the family Menispermaceae (Fig. 4.1). For all structures, SMILES (Supplementary Material) codes were used as input data in Marvin 16.1.11.0, 2016, ChemAxon (http://www.chemaxon.com). We used the software Standardizer [JChem 16.1.11.0, 2016, ChemAxon (http://www.chemaxon.com)] to perform canonicalization of the structures, add hydrogens, standardize aromaticity clean the molecular graph in three dimensions, and save compounds in *.sdf format (Imre et al., 2003).

4.2.2. VolSurf Descriptors

Three-dimensional (3D) structures were used as input data in the program VolSurf + v. 1.0.7 and were subjected to molecular interaction fields (MIFs) to generate descriptors using the following probes: N1 (amide nitrogen−hydrogen bond donor probe), O (carbonyl oxygen−hydrogen bond acceptor probe), OH_2 (water probe), and DRY (hydrophobic probe). Additional non-MIF-derived descriptors were generated to create a total of 128 descriptors (Cruciane et al., 2000). VolSurf descriptors have been previously used to predict the activity of glycogen synthase kinase-3β inhibitors (Ermondi et al., 2011) and the inhibitory profile of alkaloids against human acetylcholinesterase (Scotti and Scotti, 2015).

4.2.3. Models

The software KNIME 3.1.0 [KNIME 3.1.0 the Konstanz Information Miner, copyright, 2003−2015 (www.knime.org),] (Berthold et al., 2009) was used to perform the following analyses. The descriptors and class variables were imported from the program VolSurf + v. 1.0.7, and the data were divided using the "Partitioning" mode with the "stratified sample" option to create a training set and a test set containing 80% and 20% of the compounds, respectively. Although the compounds were randomly selected, the same proportion of active and inactive samples was maintained in both sets. For internal validation, we employed cross-validation using 10 randomly selected stratified groups, and the distributions according to activity class variables were maintained in all validation groups and in the training set. Descriptors were selected, and a model was generated using the training set and the Random Forest (RF) algorithm (Breiman, 2001) using the WEKA nodes (Hall et al., 2009). The parameters used by the RF algorithm included the following

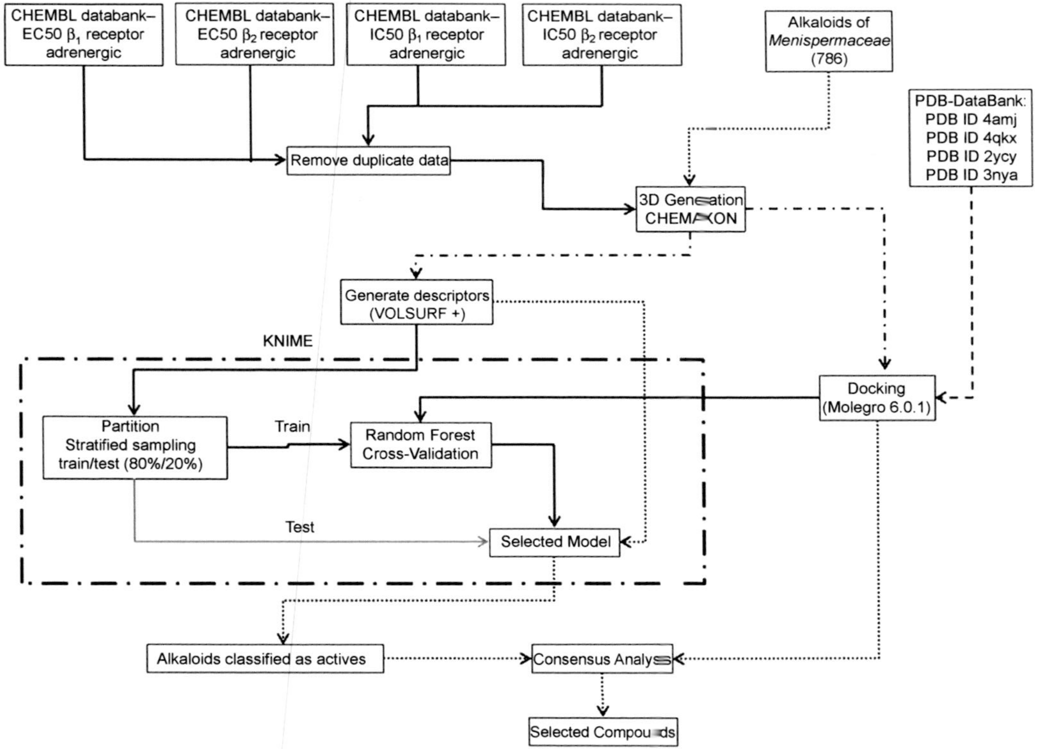

FIGURE 4.1 Virtual screening methodology used in this study. Solid black lines represent the three sets of compounds used to generate the Random Forest model for EC_{50} β_1-receptor adrenergic, EC_{50} β_2-receptor adrenergic, IC_{50} β_1-receptor adrenergic, and IC_{50} β_2-receptor adrenergic and to validate them (solid gray line: external test set). The black dotted lines represent the 786 from the alkaloids of Menispermaceae. The black dashed-dotted line represents both data sets (CHEMBL). The black dashed line represents the five enzyme structures from the PDB databank (4amj, 4qkx, 2ycy, 3nay). The dashed-dotted border delimits the process performed in the KNIME software.

setting: number of trees to build = 100. The internal and external performances of the selected models were analyzed by calculating the sensitivity (true-positive rate, i.e., active rate), specificity (true-negative rate, i.e., inactive rate), and accuracy (overall predictability). In addition, the areas under the receiver operating characteristic (ROC) curves were employed to describe true performance with more clarity than accuracy. The plotted ROC curves show the true-positive (active) rate, either versus the false-positive rates or versus the sensitivity (1-specificity). In a two-class classification, when a variable under investigation cannot distinguish between the two groups (i.e., when there is no difference between the two distributions), the area under the ROC curve equals 0.5, which is to say that the ROC curve will coincide with the diagonal. When there is a perfect separation of values between the two groups (i.e., no overlapping of distributions), the area under the ROC curve equals 1, which is to say that the ROC curve will reach the upper left corner of the plot (Hanley and McNeil, 1982).

The Matthews correlation coefficient (MCC) was calculated for training, cross-validation, and test sets, with *TP* as true-positive rate, *TN* as true-negative rate, *FP* as false-positive rate, and *FN* as false-negative rate.

$$MCC = \frac{TP \times TN - FP \times FN}{\sqrt{(TP + FP)(TP + FN)(TN + FP)(TN + FN)}}$$

4.2.4. Docking

The receptor—agonist complexes of the β_1-adrenergic receptor (PDB ID 4amj) (Warne et al., 2012), β_2-adrenergic receptor (PDB ID 4qkx) (Weicher et al., 2014), and receptor—antagonist complexes β_1-adrenergic receptor (PDB ID 2ycy) (Moukhametzianov et al., 2011) and β_2-adrenergic receptor (PDB ID 3nya) (Wacker et al., 2010) were downloaded from the Protein Data Bank (http://www.rcsb.org/pdb/home/home.do). Docking was validated by redocking the original ligand agonists carvedilol and Nb6B9 and ligand antagonists cyanopindolol and alprenolol. Alkaloid structures were submitted to molecular docking using the Molegro Virtual Docker (MVD), v. 6.0.1. All the water molecules were deleted from the enzyme structure, and the enzyme and compound structures were prepared using the same default parameter settings in the same software package. The docking procedure was performed using a GRID of 15 Å radius and 0.30 Å resolution to cover the ligand-binding site of the enzyme's structure. Template docking was used to focus the search on positions similar to the ligand interactions and conformation. The MolDock scoring algorithm was used as the score function (De Sales et al., 2015). To calculate the structure-based probability (ρ_s), we propose Eq. (4.1), where E_i, energy docking of compound i, and i varies from 1 to 63 (secondary metabolites data set); $E_{20\%}$, value power limit 20th percentile; and E_L, lower energy value of the data set.

$$\rho_s = \frac{(E_i/E_{20\%})}{-E_L} \times 100 \tag{4.1}$$

From these structures, we selected only the compounds that were classified as active when indicated by the consensus probability (ρ_c). Therefore, our methodology was to apply two screening approaches simultaneously: ligand-based screening, using the RF model based on VolSurf descriptors from four training data sets, and structure-based screening using β_1- and β_2-adrenergic receptors. To carry out this calculation, we propose Eq. (4.2), with ρ_s as structure-based probability; *TN* as true-negative rate; with ρ_l as ligand-based probability.

$$\rho_c = \frac{\rho_s + (1 + TN) \times \rho_l}{(2 + TN)} \tag{4.2}$$

Alkaloids that would be multitarget ligands against β_1- and β_2-adrenergic receptors, and those that satisfied the activity probability of EC_{50} β_2-receptor >0.6, IC_{50} β_1-receptor >0.6, $\mu_1 > 1$, and $\mu_2 > 1$, were considered where

$$\mu_1 = \frac{EC_{50}}{IC_{50}} \; \beta_2 \text{ receptor,}$$

$$\mu_2 = \frac{IC_{50}}{EC_{50}} \; \beta_1 \text{ receptor.}$$

4.3. RESULTS AND DISCUSSION

The program VolSurf v 1.0.7 generated 128 descriptors, which together with the categorical variable describe each compound as active (A) or inactive (I) using as input data the software KNIME v. 3.1.0, and generated a model RF. For all 1144 compounds, it was found that 342 had EC_{50} β_1-adrenergic receptor activity values, 99 had EC_{50} β_2-adrenergic receptor activity values, 374 had IC_{50} β_1-adrenergic receptor activity values, and 329 had IC_{50} β_2-adrenergic receptor activity values, which took approximately 3 h, using a computer with an i7 processor, 3.6 GHz and 16 GB RAM.

Table 4.1 summarizes the statistical indices of the RF models for the training, cross-validation, and test sets. For the training set, the learning machine program gave good hit rates for the inactive and active compounds, with 100% accuracy for all receptors. The RF model for EC_{50} β_2-adrenergic receptor activity showed specificity for cross-validation and test sets of 69% and 83%, respectively. The specificity results for IC_{50} β_1-adrenergic receptor activity for cross-validation and test sets were 73% and 70%, respectively

TABLE 4.1 Summary of Training, Internal Cross-Validation, Test Results, and Corresponding Match Results, Which Were Obtained Using the RF Algorithm on the Total Set of 342, 99, 374, and 329 Compounds of EC_{50} β_1-Receptor Adrenergic, EC_{50} β_2-Receptor Adrenergic, IC_{50} β_1-Receptor Adrenergic, and IC_{50} β_2-Receptor Adrenergic, Respectively

	Training			Validation		Test		
	Samples	Match	%Match	Match	%Match	Samples	Match	%Match
EC_{50} β_1-Receptor Adrenergic								
Active	135	135	100	97	72	34	27	79
Inactive	139	139	100	97	70	34	28	82
Overall	274	274	100	194	71	68	55	81
EC_{50} β_2-Receptor Adrenergic								
Active	54	54	100	50	93	13	11	85
Inactive	26	26	100	18	69	6	5	83
Overall	80	80	100	68	85	19	16	84
IC_{50} β_1-Receptor Adrenergic								
Active	143	143	100	109	76	35	23	66
Inactive	156	156	100	119	76	39	29	74
Overall	299	299	100	228	76	74	52	70
IC_{50} β_2-Receptor Adrenergic								
Active	152	152	100	124	82	38	25	66
Inactive	111	111	100	81	73	27	19	70
Overall	263	263	100	205	78	65	44	68

RF, random forest.

(Table 4.1). On the other hand, the highest values of sensitivity for cross-validation and test sets are 82% and 66%, generated by the RF model for IC_{50} β_2-adrenergic receptor activity, and 72% and 79% for the RF model of EC_{50} β_1-adrenergic receptor activity, respectively (Table 4.1).

The ROC plot generated for the test set, plotting the true-positive (active compounds) rate (sensitivity) against false-positive rates (1 − specificity), shows a greater area for EC_{50} β_2-adrenergic receptor (0.88) (Fig. 4.2B) and lower area for IC_{50} β_1-adrenergic receptor (0.73) (Fig. 4.2C).

The MCC values of training, cross-validation, and test sets are shown in Table 4.2. An MCC value of 1 represents a perfect prediction, 0 represents random prediction, and −1 represents total disagreement between prediction and observation. The MCC values for training were 1 for all the enzymes. In the test sets, the lowest value of MCC presented was 0.36 (IC_{50} β_2-adrenergic receptor) and the highest value was 0.68 (EC_{50} β_1- and β_2-adrenergic receptors). For the cross-validation, the lowest value obtained was 0.42 (EC_{50} β_1-adrenergic receptor) and the highest value was 0.64 (EC_{50} β_2-adrenergic receptor).

Docking was carried out at the active site of β_1- and β_2-adrenergic receptors using their respective agonists, carvedilol and Nb6B9, and ligand antagonists, cyanopindolol and alprenolol, respectively. For validation of the structure-based virtual screening, redocking with ligands extracted from the PDB files was performed, and the positions with the lowest energies were selected. For both β_1-and β_2-adrenergic receptors, the lowest energy positions of all ligands showed a perfect superposition with the corresponding results obtained in crystallography.

The probability of a compound being potentially active was obtained from the combined probabilities of the structure-based virtual screening (docking) and ligand-based virtual screening (RF models) that can be seen in Fig. 4.1 and Eqs. (4.1) and (4.2), which represent the consensus analysis proposed in this study.

We evaluated the potential of the alkaloids present in the family Menispermaceae using structure-based virtual screening (docking) and ligand-based virtual screening (RF models) on selected enzyme targets that are involved in the inflammatory process. The results showed that five compounds: (1*R*,9*S*,10*S*)-3-hydroxy-4,11,12-trimethoxy-17-methyl-17-azatetracyclo[7.5.3.$0^{1,10}$.$0^{2,7}$]heptadeca-2,4,6,11-tetraen-13-one (**9**); (1*S*,9 *R*,10*R*)-3,12-dihydroxy-4,11-dimethoxy-17-methyl-17-azatetracyclo[7.5.3.$0^{1,10}$.$0^{2,7}$]heptadeca-2(7),3,5,11-tetraen-13-one (**154**); **337**; 6,14,15,16-tetramethoxy-10-azatetracyclo[7.7.1.$0^{2,8}$.$0^{13,17}$]heptadeca-1(17),2(8),3,6,9,11,13,15-octaen-5-one (**476**); and (−)-reticuline ((1*R*)-1-[(3-hydroxy-4-methoxyphenyl)methyl]-6-methoxy-2-methyl-1,2,3,4-tetrahydroisoquinolin-7-ol) (**625**) presented a binding selectivity for β_2-adrenergic receptor greater than 80% (Supplementary Material). Research shows that (*S*)-reticuline has effects on vascular smooth muscle cells through the inhibition of Ca^{2+} channels (Medeiros et al., 2008). In another study on

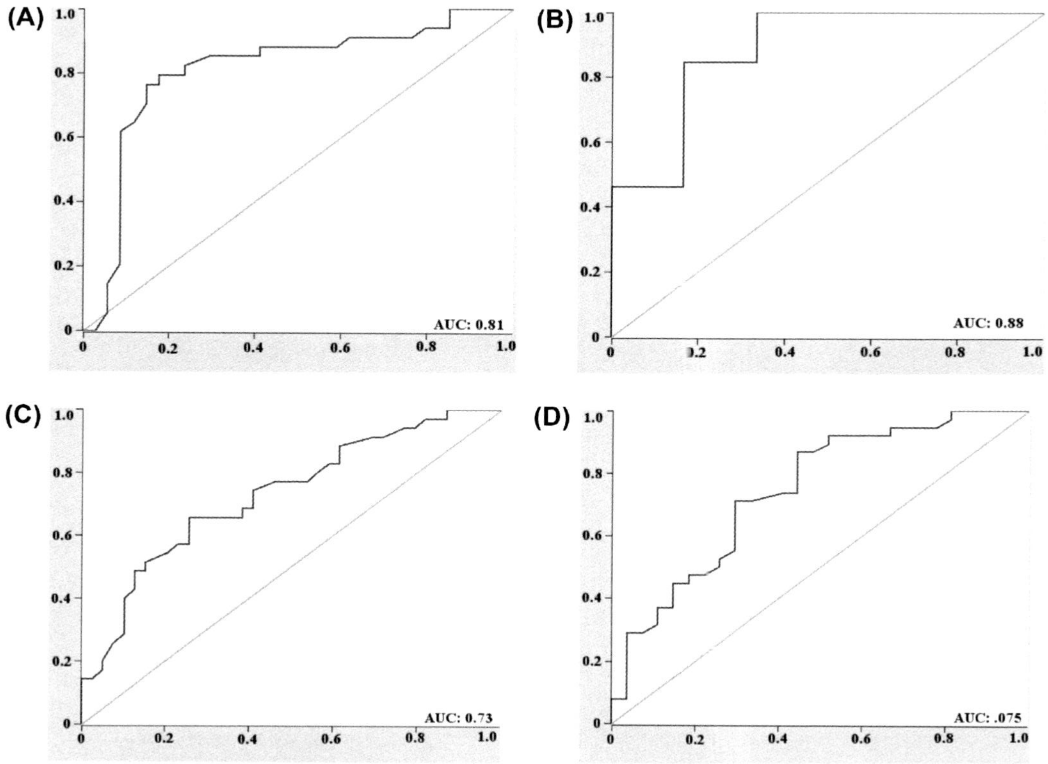

FIGURE 4.2 Receiver operating characteristic plot generated by the selected RF models for the test set and value of the area under the curve (AUC). (A) EC_{50} β_1-receptor adrenergic, (B) EC_{50} β_2-receptor adrenergic, (C) IC_{50} β_1-receptor adrenergic, and (D) IC_{50} 3_2-receptor adrenergic.

TABLE 4.2 MCC Values Of Training, Cross-Validation, and Test Sets for the Three Models: β_1- and β_2-Receptor Adrenergic

Model	Validation	Test
EC_{50} β_1-receptor adrenergic	0.42	0.68
EC_{50} β_2-receptor adrenergic	0.64	0.68
IC_{50} β_1-receptor adrenergic	0.53	0.40
IC_{50} β_2-receptor adrenergic	0.55	0.36

MCC, Matthews correlation coefficient.

reticuline, as tested in normotensive rats, it was found to cause a hypotensive effect by inhibiting the release of Ca^{2+}, and the Ca^{2+} channel locked the dependent voltage (Dias et al., 2004).

For the binding selectivity of β_1-adrenergic receptor, virtual screening (VS) selected four compounds with a percentage higher than 65%: (1*S*)-1-({3-[(4-{[(1*S*)-6,7-dimethoxy-2-methyl-3,4-dihydro-1H-isoquinolin-1-yl]methyl}phenyl) methoxy]-4-hydroxyphenyl}methyl)-6-methoxy-1,2,3,4-tetrahydroisoquinolin-7-ol (**1**); (1*R*)-1-({4-hydroxy-3-[(4-{[(1*R*)-7-hydroxy-6-methoxy-1,2,3,4-tetrahydroisoquinolin-1-yl]methyl}phenyl)methyl]phenyl}methyl)-6-methoxy-2-methyl-3,4-dihydro-1H-isoquinolin-7-ol (**147**); (*S*)-*N*-methylcoclaurine ((1*S*)-1-[(4-hydroxyphenyl)methyl]-6-methoxy-2-methyl-1,2,3,4-tetrahydroisoquinolin-7-ol (**393**); and (12b*S*)-2,10-dimethoxy-7,8,12b,13-tetrahydro-5H-6-azatetraphene-3,11-diol (**692**) (Supplementary Material).

The results presented in Table 4.3 show the compounds selected by the consensus analyses, with potential for selective multitarget connection for both β_2-adrenergic agonists and β_1-adrenergic antagonists receptors.

From the 786 alkaloids, the compounds coclaurine ((1*S*)-1-[(4-hydroxyphenyl) methyl]-6-methoxy-1,2,3,4-tetrahydroisoquinolin-7-ol) (**285**); (3*S*,11a*R*)-3, 8,9-trimethoxy-2,3,4,6,11-octahydro-5a-azatetraphen-6-one (**360**); (1*S*)-1-[(4-hydroxyphenyl)methyl]-7-methoxy-5,8-dimethyl-1,2,3,4-tetrahydroisoquinolin-6-ol (**374**); macoline ((1*S*,14 *R*)-6,21-dihydroxy-20,25-dimethoxy-15,15,30-trimethyl-8,23-dioxa-15,30-diazaheptacyclo[22.6.2.$2^{9,12}$.$1^{3,7}$.$1^{14,18}$.$0^{27,31}$.$0^{22,33}$]hexatriaconta-3(36),4,6,9,11,18,20,22(33),24,26,31,34-dodecaen-15-ium) (**448**); dauricine (4-{[(1*R*)-6,7-dimethoxy-2-methyl-1,2,3,4-tetrahydroisoquinolin-1-yl]methyl}-2-(4-{[(1*R*)-6,7-dimethoxy-2-methyl-1,2,3,4-tetrahydroisoquinolin-1-yl]methyl} phenoxy)phenol (**482**); and 4-{[(1*S*)-6,7-dimethoxy-1,2,3,4-tetrahydrois oquinolin-1-yl]methyl}phenol (**703**) were selected for these receptors (Table 4.3). It can be seen from the values of structure-based and ligand-based VS that these six alkaloids showed higher EC_{50} and IC_{50} in the receptors under study (Fig. 4.3). The alkaloids **285**, **360**, **374,** and **448** have methoxy

TABLE 4.3 Indication of the Compounds' Considerable activity Through a Ligand-Based Approach in Each Receptor, and the Percentage of Prediction as active (P. A. %)

	EC_{50} β_1-Receptor Adrenergic			EC_{50} β_2-Receptor Adrenergic			IC_{50} β_1-Receptor Adr energic			IC_{50} β_2-receptor Adrenergic			Multitarget	
ID	ρ_s(%)	ρ_s(%)	ρ_s(%)	ρ_s(%)	ρ_s(%)	ρ_s(%)	ρ_s(%)	ρ_s(%)	ρ_s(%)	ρ_s(%)	ρ_s(%)	ρ_s(%)	μ_1	μ_2
285	38	73	60	47	88	73	39	78	64	49	80	69	1.06	1.06
360	36	74	60	49	78	67	39	74	61	46	76	65	1.03	1.02
374	35	69	56	47	82	69	34	75	60	44	79	66	1.04	1.06
448	42	49	47	42	75	63	50	68	61	28	72	56	1.12	1.31
482	51	44	47	70	61	64	74	53	61	73	58	64	1.01	1.30
703	38	75	61	49	82	70	41	77	64	47	79	67	1.03	1.04

Compounds selected as active and inactive receptors are given in bold.

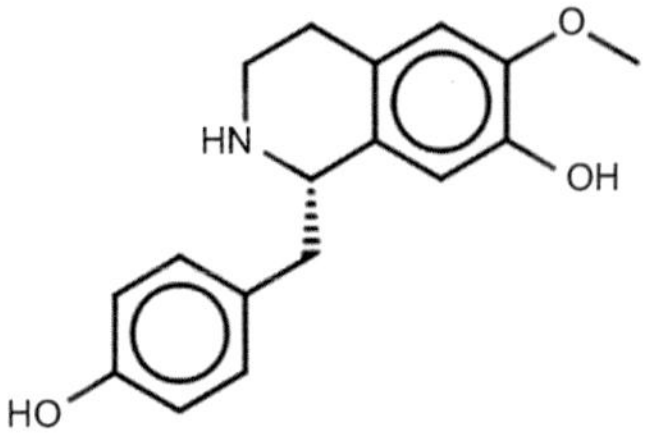

(285)

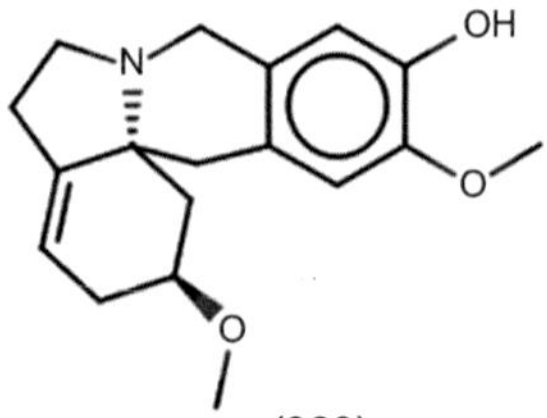

(360)

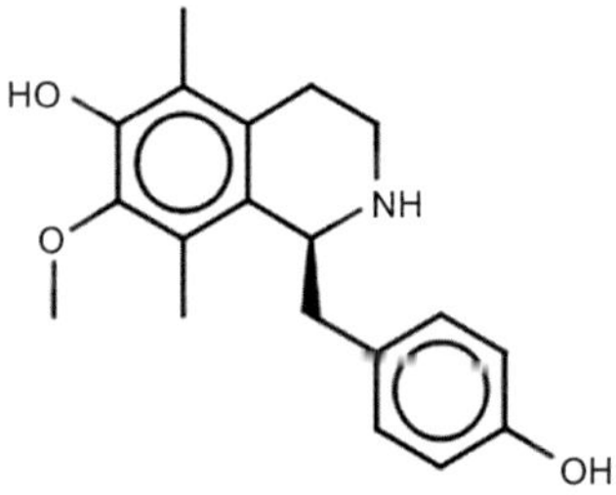

(374)

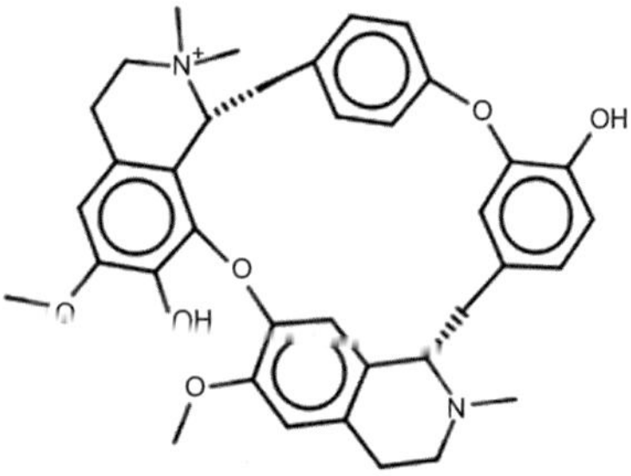

(448)

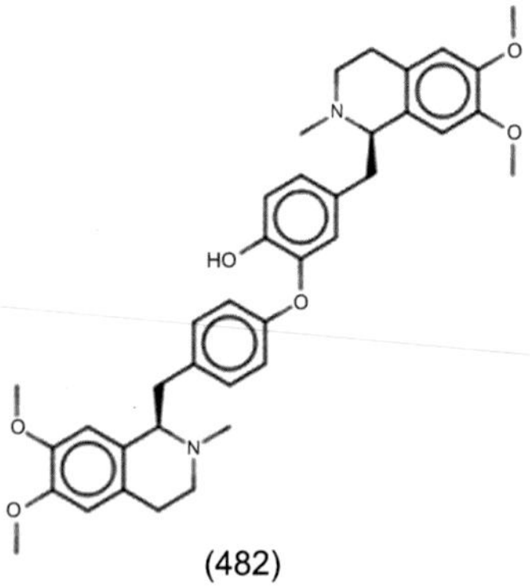

(482)

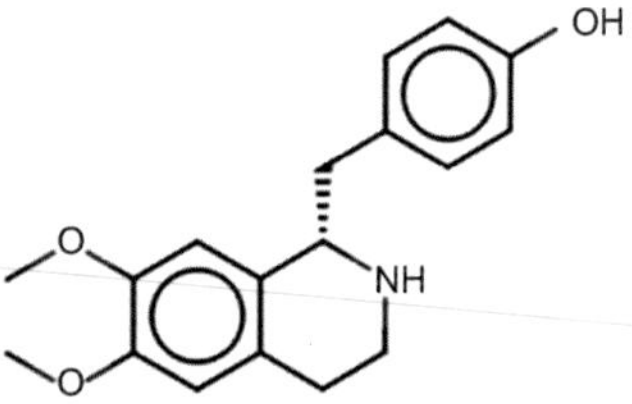

(703)

FIGURE 4.3 Chemical structures of the multitarget compounds.

and hydroxy groups in positions 6 and 7, respectively, in their chemical structure. The alkaloids **482** and **703** have two methoxy groups in positions 6 and seven in their structures. One can also note that there is a similarity between the alkaloids **285**, **374,** and **703**, where we find the presence of the hydroxybenzyl group.

Compared with the other five alkaloids, the cyclic compound named macoline (**448**) proved to be the one with the greatest selectivity for binding the activating β_2-receptor and the inhibiting receptor β_1 in a multitarget

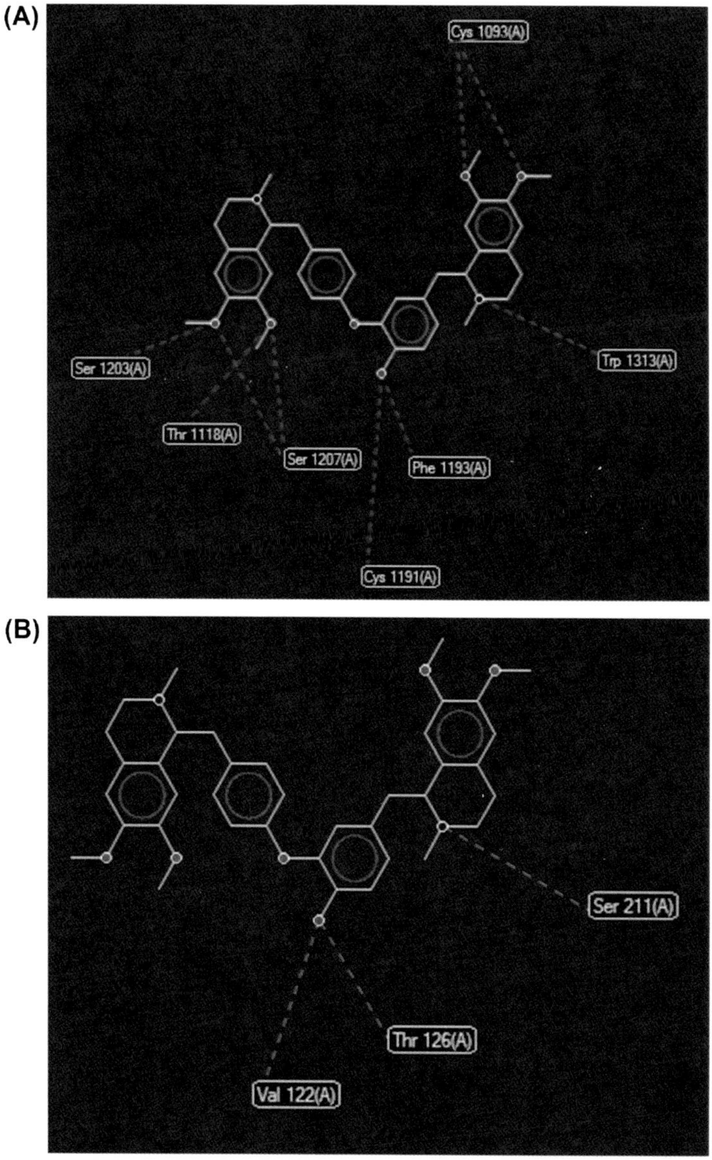

FIGURE 4.4 Docking carried out with the alkaloid dauricine (**482**), which has improved bonding power with antagonist β_1-receptor adrenergic (A) and agonist β_2-receptor adrenergic (B).

analysis. Alkaloid dauricine (**482**) showed better MolDock energies for both models antagonist β_1-receptors and agonist β_2-adrenergic receptors (Fig. 4.4). This compound shows a hydrogen interaction between the receptors in this study, the interactions at the β_2-adrenergic receptor being stronger (Fig. 4.4B).

The antiarrhythmic effects of dauricine have been the subject of several studies. Researches show that dauricine acts by inhibiting calcium channels in cardiomyocytes and inhibiting potassium current in ventricular myocytes (Liu et al., 2009; Xia et al., 2000). This alkaloid has been tested for human-Ether-à-go-go-Related Gene (hERG) potassium channels, which could inhibit it by inducing a marked change of hERG activation curves toward negative potentials (Zhao et al., 2012).

4.4. CONCLUSION

In the present study, we evaluated 786 alkaloids belonging to the family Menispermaceae through two different methodologies, ligand-based virtual screening using the RF algorithm and VolSurf descriptors, and a structure-based approach using MVD, in order to determine the potential for activation in β_2-adrenergic receptors and inhibition in β_1-adrenergic receptors important for asthma and heart disease, respectively. From the 786 alkaloids, six of them were selected as potential multitarget ligands for the aforementioned receptors. Compound **482** was the only one that had no potential activity and inhibited the receptors under study. These compounds can be used as starting points for further studies to propose structures for the treatment of various diseases.

ACKNOWLEDGMENTS

We are grateful to the -development agency CNPq.

Conflict of Interest: The authors declare no conflicts of interest.

REFERENCES

Berthold, M.R., Cebron, N., Dill, F., Gabriel, T.R., Kötter, T., Meinl, T., Ohl, P., Thiel, K., Wiswedel, B., 2009. KNIME—the Konstanz information miner. ACM SIGKDD Explor. Newsl. 11 (1), 26.

Bezerra-Santos, C.R., Vieira-de-Abreu, A., Vieira, G.C., Filho, J.R., Barbosa-Filho, J.M., Pires, A.L., Martins, M.A., Souza, H.S., Bandeira-Melo, C., Bozza, P.T., Piuvezam, M.R., 2012. Effectiveness of *Cissampelos sympodialis* and its isolated alkaloid warifteine in airway hyperreactivity and lung remodeling in a mouse model of asthma. Int. Immunopharmacol. 13 (2), 148—155.

Billingsley, M.L., 2008. Druggable targets and targeted drugs: enhancing the development of new therapeutics. Pharmacology 82 (4), 239—244.

Breiman, L., 2001. Random forests. Mach. Learn. 45 (1), 5—32.

Cavalcante, H.M.M., Ribeiro, T.P., Silva, D.F., Nunes, X.P., Barbosa-Filho, J.M., Diniz, M.F.F.M., Correia, N.A., Braga, V.A., Medeiros, I.A., 2010. Cardiovascular effects elicited by milonine, a new 8, 14-dihydromorphinandienone alkaloid. Basic Clin. Pharmacol. Toxicol. 108 (2), 122—130.

Cruciani, G., Crivori, P., Carrupt, P., Testa, B., 2000. Molecular fields in quantitative structure—permeation relationships: the VolSurf approach. J. Mol. Struct.: THEOCHEM 503 (1—2), 17—30.

De Sales, I.R.P., Machado, F.D.F., Marinho, A.F., Lúcio, A.S.S.C., Filho, J.M.B., Batista, L.M., 2015. *Cissampelos sympodialis* Eichl. (Menispermaceae), a medicinal plant, presents antimotility and antidiarrheal activity *in vivo*. BMC Complement. Altern. Med. 15 (1).

Dias, K.L., Da Silva, D.C., Barbosa-Filho, J.M., Almeida, R.N., De Azevedo, C.N., Medeiros, I.A., 2004. Cardiovascular effects induced by reticuline in normotensive rats. Planta Med. 70 (4), 328–333.

Ermondi, G., Caron, G., Pintos, I.G., Gerbaldo, M., Pérez, M., Pérez, D.I., Gándara, Z., Martínez, A., Gómez, G., Fall, Y., 2011. An application of two MIFs-based tools (Volsurf+ and Pentacle) to binary QSAR: the case of a palinurin-related data set of non-ATP competitive glycogen synthase kinase 3β (GSK-3β) inhibitors. Eur. J. Med. Chem. 46 (3), 860–869.

Fitzgerald, M.K., Otis, J.M., Mueller, D., 2016. Dissociation of β1- and β2-adrenergic receptor subtypes in the retrieval of cocaine-associated memory. Behav. Brain Res. 296, 94–99.

Hagg, A., Colgan, T.D., Thomson, R.E., Qian, H., Lynch, G.S., Gregorevic, P., 2016. Using AAV vectors expressing the β_2-adrenoceptor or associated Gα proteins to modulate skeletal muscle mass and muscle fibre size. Sci. Rep. 6, 23042.

Hall, M., Frank, E., Holmes, G., Pfahringer, B., Reutemann, P., Witten, I.H., 2009. The WEKA data mining software. ACM SIGKDD Explor. Newsl. 11 (1), 10.

Hanley, J.A., McNeil, B.J., 1982. The meaning and use of the area under a receiver operating characteristic (ROC) curve. Radiology 143 (1), 29–36.

Hoffmann, C., Leitz, M.R., Oberdorf-Maass, S., Lohse, M.J., Klotz, K., 2004. Comparative pharmacology of human β-adrenergic receptor subtypes?Characterization of stably transfected receptors in CHO cells. Naunyn-Schmiedeberg's Arch Pharmacol 369 (2), 151–159.

Imre, G., Veress, G., Volford, A., Farkas, Ö., 2003. Molecules from the Minkowski space: an approach to building 3D molecular structures. J. Mol. Struct.: THEOCHEM 666-667, 51–59.

Liu, Q.-N., Zhang, L., Gong, P.-L., Yang, X.-Y., Zeng, F.-D., 2009. Inhibitory effects of dauricine on early afterdepolarizations and L-type calcium current. Can. J. Physiol. Pharmacol. 87 (11), 954–962.

Medeiros, M.A.A., Nunes, X.P., Barbosa-Filho, J.M., Lemos, V.S., Pinho, J.F., Roman-Campos, D., de Medeiros, I.A., Araújo, D.A.M., Cruz, J.S., 2008. (*S*)-reticuline induces vasorelaxation through the blockade of L-type Ca^{2+} channels. Naunyn-Schmiedeberg's Arch. Pharmacol. 379 (2), 115–125.

Moukhametzianov, R., Warne, T., Edwards, P.C., Serrano-Vega, M.J., Leslie, A.G.W., Tate, C.G., Schertler, G.F.X., 2011. Two distinct conformations of helix 6 observed in antagonist-bound structures of a β_1-adrenergic receptor. Proc. Natl. Acad. Sci. U. S. A. 108 (20), 8228–8232.

Peitzman, E.R., Zaidman, N.A., Maniak, P.J., O'Grady, S.M., 2015. Carvedilol binding to β2 -adrenergic receptors inhibits CFTR-dependent anion secretion in airway epithelial cells. Am. J. Physiol. Lung Cell. Mol. Physiol. 310 (1), L50–L58.

Pelaia, G., Muzzio, C.C., Vatrella, A., Maselli, R., Magnoni, M.S., Rizzi, A., 2015. Pharmacological basis and scientific rationale underlying the targeted use of inhaled corticosteroid/long-acting β2 -adrenergic agonist combinations in chronic obstructive pulmonary disease treatment. Expert Opin. Pharmacother. 16 (13), 2009–2021.

Schofield, M.L., 2014. Asthma pharmacotherapy. Otolaryngol. Clin. North Am. 47 (1), 55–64.

Scotti, L., Scotti, M., 2015. Computer aided drug design studies in the discovery of secondary metabolites targeted against age-related neurodegenerative diseases. Curr. Top. Med. Chem. 15 (21), 2239–2252.

Semwal, D.K., Semwal, R.B., Vermaak, I., Viljoen, A., 2014. From arrow poison to herbal medicine – the ethnobotanical, phytochemical and pharmacological significance of *Cissampelos* (Menispermaceae). J. Ethnopharmacol. 155 (2), 1011–1028.

Velema, W.A., Szymanski, W., Feringa, B.L., 2014. Photopharmacology: beyond proof of principle. J. Am. Chem. Soc. 136 (6), 2178–2191.

Vieira, G.C., De Lima, J.F., De Figueiredo, R.C.B.Q., Mascarenhas, S.R., Bezerra-Santos, C.R., Piuvezam, M.R., 2012. Inhaled *Cissampelos sympodialis* down-regulates airway allergic reaction by reducing lung $CD3^+$ T cells. Phytother. Res. 27 (6), 916–925.

Voigt, N., Heijman, J., Dobrev, D., 2014. Kardiovaskuläre pharmakotherapie. Herz 39 (2), 227–240.

Von Homeyer, P., Schwinn, D.A., 2011. Pharmacogenomics of β-adrenergic receptor physiology and response to β-blockade. Anesth. Analg. 113 (6), 1305–1318.

Wacker, D., Fenalti, G., Brown, M.A., Katritch, V., Abagyan, R., Cherezov, V., Stevens, R.C., 2010. Conserved binding mode of human β_2 adrenergic receptor inverse agonists and antagonist revealed by X-ray crystallography. J. Am. Chem. Soc. 132 (33), 11443–11445.

Warne, T., Edwards, P.C., Leslie, A.G.W., Tate, C.G., 2012. Crystal structures of a stabilized β_1-adrenoceptor bound to the biased agonists bucindolol and carvedilol. Structure 20 (5), 841–849.

Wefferling, K.M., Hoot, S.B., Neves, S.S., 2013. Phylogeny and fruit evolution in Menispermaceae. Am. J. Bot. 100 (5), 883–905.

Wehr, M.C., Rossner, M.J., 2016. Split protein biosensor assays in molecular pharmacological studies. Drug Discov. Today 21 (3), 415–429.

Weichert, D., Kruse, A.C., Manglik, A., Hiller, C., Zhang, C., Hubner, H., Kobilka, B.K., Gmeiner, P., 2014. Covalent agonists for studying G protein-coupled receptor activation. Proc. Natl. Acad. Sci. U. S. A. 111 (29), 10744–10748.

Wu, M., Zhang, Y., Zhou, Q., Xiong, J., Dong, Y., Yan, C., 2016. Higenamine protects ischemia/reperfusion induced cardiac injury and myocyte apoptosis through activation of β_2-AR/PI3K/AKT signaling pathway. Pharmacol. Res. 104, 115–123.

Xia, J.S., Guo, D.L., Zhang, Y., Zhou, Z.N., Zeng, F.D., Hu, C.J., 2000. Inhibitory effects of dauricine on potassium currents in Guinea pig ventricular myocytes. Acta Pharmacol. Sin. 21 (1), 60–64.

Zhao, J., Lian, Y., Lu, C., Jing, L., Yuan, H., Peng, S., 2012. Inhibitory effects of a bisbenzylisoquinline alkaloid dauricine on HERG potassium channels. J. Ethnopharmacol. 141 (2), 685–691.

Chapter 5

Natural Chemotherapeutic Agents for Cancer

R. Dutt[1], V. Garg[2], A.K. Madan[3]
[1]*G.D. Goenka University, Gurgaon, India;* [2]*Maharshi Dayanand University, Rohtak, India;* [3]*Pt. B.D. Sharma University of Health Sciences, Rohtak, India*

5.1. INTRODUCTION

Despite the significant technological advances in the comprehension of neoplastic diseases during the past few decades, cancer is the third leading cause of death globally, accounting for approximately 8 million deaths and 13 million new cases per year (Tímár et al., 2001; Dal Lago et al., 2008; Rebecca et al., 2012). Such a high mortality rate itself highlights the limitation of conventional treatment of cancer including radiotherapy, chemotherapy, and surgery in controlling this life-threatening disease (Talib, 2011). The prevalence of this dreadful human disease is increasing with modern life style, food habits, and environmental factors. Although numerous synthetic drugs with high potency have been developed for chemotherapy of cancer during past five decades, their toxicity to normal cells as well as drug resistance have emerged as the major obstacles for their clinical success.

Major categories of chemotherapeutic agents are cytotoxic in nature, which decreases considerably the quality of life of patients and affects the life of survivors for a considerable period after the treatment. Toxicity often limits the usefulness of anticancer agents. Toxicity is also the major cause for discontinuation of treatment by many patients. The surgical procedure cannot be applied once cancer has spread. Moreover, radiotherapy and chemotherapy are not cancer cell specific and can induce life-threatening adverse effects. Therefore, these anticancer therapies have significant limitations and cannot cope up with the increasing cancer incidence (Wang et al., 2006). Anticancer products derived from nature definitely offer invaluable means to raise cancer therapy to a new level of success with possibly less or no adverse side effects. Alternatively, numerous natural products have been reported to improve the efficiency of chemotherapeutic agents with less or no side effects. The precise use of phytochemicals in cancer chemotherapy may provide new dimensions

Multi-Scale Approaches in Drug Discovery. http://dx.doi.org/10.1016/B978-0-08-101129-4.00005-9

for improving its outcome in a complementary way (Sak, 2012), and discovery of new anticancer agents derived from nature, especially plants, is currently under investigation. As per estimates of the World Health Organization, global market of herbal medicines is approximately US $83 billion annually (Robinson and Zhang, 2011; Rivera et al., 2013). A major proportion of clinically approved chemotherapeutic agents were either pure phytoconstituents or their semisynthetic derivatives, or synthesized molecules based on natural lead molecule (Newman and Cragg, 2007). Several of them are currently in various phases of clinical trials or undergoing further investigation (Corson and Crews, 2007; Butler, 2008; Harvey, 2008; Saklani and Kutty, 2008; Sashidhara et al., 2009; Pan et al., 2013).

5.2. PLANTS AS A SOURCE OF CHEMOTHERAPEUTIC AGENTS

Natural products, particularly those from plant kingdom have been an important source of chemotherapeutics for the past 30 years. More than half of the effective anticancer drugs can be traced to herbal origin (Max and Wang, 2009). For a long time, plants have been used as an important therapeutic source for the treatment of cancer as herbal teas or juices, as crude extracts, or as standard enriched fractions in pharmaceutical preparations such as tinctures, fluid extracts, powders, tablets, and capsules (Karakas et al., 2012). Most chemotherapeutic agents have been developed through random screening of organism collections. With the current advances in understanding of carcinogenesis it is now possible to develop more and more efficient strategies to specifically explore ancient species for the isolation of chemotherapeutic agents. Shortlisting of plants growing in diverse hostile environments may be considered as a vital screening strategy for the accelerated development of new chemotherapeutic agents for cancer (Dutt et al., 2014). "*Cancer is hostile to human body and hostile en vironment is cancerous to plants,*" Accordingly, only those plants that can easily grow in multiple but diverse hostile environments may naturally provide a vital source of anticancer agents. This strategy is of utmost significance because of the ease in shortlisting the vast flora (2,50,000–5,00,000 plant species) into very few plants for phytochemical and biological screening (Dutt et al., 2014). Such strategies would not only enhance success rate in terms of providing effective and safe chemotherapeutics agents but also minimize the risk of postmarketing withdrawals (Katiyar et al., 2012). The pioneering work by late Dr. Jonathan Hartwell (1969) involving active phytoconstituents of *Podophyllum peltatum* L. (Berberidaceae) and isolation of vinblastine and vincristine from *Cathranthus roseus* by late Dr. Gordon Svoboda (1975) provided substantial evidence that the plant kingdom can serve as a potential source of novel chemotherapeutic agents (Roja and Rao, 2000). The initiation the screening program of plants for anticancer activity by the United States National Cancer Institute (NCI) in 1960 led to the discovery of new chemical entities of diverse scaffold showing a range of cytotoxic activities (Cassady and Douros, 1980). During the period 1986–2004, more than 60,000 plant species were collected from the various targeted tropical and subtropical regions of the world by the Natural Products Branch of the Developmental Therapeutics Program at the NCI. The plant extracts of these species were prescreened initially against a 60 cancer cell lines representing nine cancer types (Pan et al., 2010; Cragg et al., 1999, 2006). The outcome of these screening programs resulted in the development of the most exciting plant-derived anticancer drug—*taxol*, which is now named *paclitaxel*. This potential phytoconstituent was originally isolated from the bark of the Pacific Yew, *Taxus brevifolia* Nutt. (Taxaceae) (Wani et al., 1971). The discovery of pharmacological action of *taxol* was a key milestone in the anticancer drug discovery. It was ultimately introduced in the US market for clinical use in the early 1990s (Wall and Wani, 1996; Balunas and Kinghorn, 2005). Isolation and characterization of chemotherapeutic agents from plants continues unabated. Plants with promising chemotherapeutic agents have been enlisted in Table 5.1.

5.3. DIETARY SUPPLEMENTS IN CHEMOTHERAPY

Some dietary supplements protects the body from cancer through detoxification, while others aid in reducing the toxic side effects of chemotherapy and radiotherapy. Globally, scientists are concentrated on the role of dietary supplements of plant origin to boost immune cells of the body against cancer. Plant-based supplement therapies are gaining importance in ensuring the best health before undergoing chemotherapeutic treatment (Brom, 2009). The adverse side effects associated with chemotherapy exacerbate the nutritional problems, and the proper nutrition supplement should be an integral part of the anticancer treatment (Philips, 1999). Dietary components may act as adaptogens preventing organism from the adverse side effects through either intervening or modifying the biologic response (Sagar et al., 2006; Davis, 2007). Thus, this approach may prove beneficial in chemotherapeutic treatment by supplementing or supporting the body with plant nutrients so as to assist in overall effectiveness of chemotherapy (Osiecki, 2002). In Table 5.2, some of the reported dietary supplements have been enlisted. These dietary supplements can aid in enhancing the effect of chemotherapy with lesser side effects. Patient's own initiative in consuming these supplements may be dangerous and should be avoided because it may attenuate or inhibit the therapeutic effect of certain chemotherapeutic drugs. The patient should consult nutritional oncologist for optimal dietary regimen.

TABLE 5.1 Plants Having Potential for Anticancer Activity

Family	Name of Plants	Bioactive Constituents	References
Actinidiaceae	*Actinidia chinensis*	Triterpenes, polyphenols, and anthraquinones	Zhu et al. (2013)
Agaricaceae	*Lentinus edodes*	Olysaccharide	Chihara et al. (1970)
Amaranthaceae	*Iresine herbstii*	Sesquiterpene Iresine, tlatlancuayin, and isoflavones and carbohydrates	Dipankar et al. (2011)
Anacardiaceae	*Anacardium occidentale, Myracrodruon urundeuva*	2-Hydroxy-6-pentadecylbenzoic acid and 2,6-dihydroxybenzoic acid, myricetin, quercetin, kaempferol, rhamnetin, cyanidin, peonidin, and delphinidin	Kubo et al. (1993) and Ferreira et al. (2011)
Annonaceae	*Annona glabra*	(2E,4E,1′R,3′S,5′R,6′S)-dihydrophaseic acid, icariside D2,3,4-dimethoxyphenyl O-β-D-glucopyranoside, 3,4-dihydroxybenzoic acid, blumenol A, cucumegastigmane , and icariside B1	Cochrane et al. (2008)
Apiaceae	*Anethum graveolens, Angelica sinensis*	Polysaccharides	Zheng et al. (1992), Zhu et al. (2012), and Cao et al. (2010)
Apocynacaeae	*Hancornia speciosa, Bleekeria vitensis, Catharanthus roseus, Ervatamia coronaria, Forsteronia refracta, Himatanthus articulates, H. sucuuba, Ochrosia borbonica, O. elliptica*	Cyclitols, quinic acid, -bornesitol, ervatamine, apparicine, and coronaridine	Endringer et al. (2009), Cragg and Suffness (1988), Paoletti et al. (1980), Carter and Livingstone (1976), El-Sayed and Cordell (1981), El-Sayed et al. (1983), Hullatti et al. (2013), Xu et al. (2006a,b), Rebouças Sde et al. (2011), Persinos and Blomster (1978), Svoboda et al. (1968), and Kuo et al. (2005)

Araceae	*Arisaema tortuosum*	Quercetin, rutin, luteolin, and lectin	Dhuna et al. (2005)
Araliacea	*Aralia nudicaulis, Acanthopanax gracilistylus, Panax ginseng*	Hexane fraction of rhizome and fruit	Wang et al. (2006), Huang et al. (2006), Shan et al. (1999), Xie (1989), and Chang et al. (2003)
Arecaceae	*Orbignya phalerata*	Ethanol extract	Rennó et al. (2008)
Asclepiadaceae	*Marsdenia tenocissima*	β-sitosterol, condutirol, dihydroconduritol, betulinic acid, lupeol, and daucosterol	Zhang et al. (2010)
Asteraceae	*Acanthospermum hispidum, Aster squamatus, Artemisia glabella, Calendula officinalis, Centaurea montana, C. jacea, C. schischkinii, Echinacea angustifolia, Berberis amurensis, Eupatorium formosanum, Gynoxys verrucosa, Silybum marianum, Tanacetum parthenium, Elephantopus mollis*	Ten oleanane-type triterpene glycosides, including four new compounds, calendulaglycoside A 6′-O-methyl ester, calendulaglycoside A 6′-O-n-butyl ester, calendulaglycoside B 6′-O-n-butyl ester, and calendulaglycoside C 6′-O-n-butyl ester, cirsiliol, apigenin, hispidulin, eupatorin, isokaempferide, axillarin, centaureidin, 6-methoxykaempferol 3-methyl ether, trachelogenin, cincin, 4′-acetylcnicin leucodine, dehydroleucodine, arglablin	Rajendran and Deepa (2007), Bibi et al. (2011), Ukiya et al. (2006), Shoeb (2006), Forgo (2012), Huntimer et al. (2006), Lee et al. (1972), Habtemariam and Macpherson (2000), Tyagi et al. (2002), Parada-Turska et al. (2007), Ooi et al. (2011), Ordóñez et al. (2016), and Lone (2015)
Berberidaceae	*Berberis amurensis, Podophyllum emodi, P. peltatum*	Podophyllotoxins	Xie et al. (2009), Xu et al. (2006a,b), Park et al. (2009), Duan et al. (2010), Stähelin (1973), and Damayanthi and Lown (1998)
Betulaceae	*Betula alba, B. utilis*	Betulinic acid	Fulda (2008) and Patočka (2003)

Continued

TABLE 5.1 Plants Having Potential for Anticancer Activity—cont'd

Family	Name of Plants	Bioactive Constituents	References
Bignoniaceae	*Tabebuia avellanedae*	Lapachol, lapachol methyl ether, deoxy lapachol, β-lapachone, α-lapachone, and dehydro-α- lapachone	Rao et al. (1968), Hussain et al. (2007), and Epifano et al. (2014)
Boraginaceae	*Symphytum officinale*	Indicine-N-oxide	Roman et al. (2008)
Brassicaceae	*Brassica oleracea*	Sulforaphane	Devi and Thangam (2012)
Bromeliaceae	*Ananas comosus*	Bromelain (mixture of proteases)	Chobotova et al. (2010)
Cannabaceae	*Cannabis sativa*	Ω-9-Tetrahydrocannabinol, delta-8-tetrahydrocannabinol, and cannabinol	Munson et al. (1975), Casanova (2003), and Kerbel and Folkman (2002)
Caricaceae	*Carica papaya*	5,7-Dimethoxycoumarin	Alesiani et al. (2008) and Breemen and Pajkovic (2008)
Casealpiniaceae	*Delonix regia*	Sterols—stigmasterol; β-sitosterol and its 3-O-glucoside; a triterpene, namely, ursolic acid; flavonoids: quercetin, quercitrin, isoquercitrin, and rutin	El-Sayed et al. (2011)
Celastraceae	*Maytenus ilicifolia, Tripterygium wilfordii*	Cytotoxic aromatic triterpenes and quinoid triterpenes	Costa et al. (2008), Shirota et al. (1994), and Wong et al. (2012)
Cephalotaxacea	*Cephalotaxus harringtonia*	Cephalotaxine and harringtonine	Powell et al. (1970), Ohnuma and Holland (1985), and Beranova et al. (2013)

Chenopodiaceae	*Chenopodium ambrosioides*	Kaempferol-7-O-alpha-L-rhamnopyranoside, kaempferol-3,7-di-O-alpha-L-rhamnopyranoside, patuletin, quercetin-7-O-alpha-L-rhamnopyranoside, grasshopper ketone, 4-hydroxy-4-methyl-2-cyclohexen-1-one, syringaresinol, benzyl beta-D-glucopyranoside, dendranthemoside B, N-trans-feruloyl tyramine	Nascimento et al. (2006) and Efferth et al. (2002)
Chinese recipe	*Danggui Longhui Wan*	Indirubin	Hoessel et al. (1999)
Chlorellaceae	*Chlorella pyrenoidosa*	CPPS Ia and CPPS IIa polysaccharide	Sheng et al. (2007)
Chrysobalanaceae	*Parinari curatellifolia*	Diterpenoids (13-methoxy-15-oxozoapatlin and 13-hydroxy-15-oxozoapatlin)	Lee et al. (1996)
Colchiaceae	*Colchicum autumnale*	Colchicines	Lindholm et al. (2002) and Bhattacharyya et al. (2014)
Combretaceae	*Combretum caffrum*	Combretastatin A-1	Dorr et al. (1996)
Compositae	*Arctium lappa, Artemisia argyi*	Sesquiterpene, roots contain inulin and flavonoids, fruits contain lignin	Machado et al. (2012) and Seo et al. (2003)
Convolvulaceae	*Ipomoea batatas*	4-Ipomeanol	Christian et al. (1989) and da Rocha et al. (2001)
Cucurbitaceae	*Cucurbita andreana*	Cucurbitacins B, D, E, and I	Jayaprakasam et al. (2003) and Bernard and Olayinka (2010)
Dioscoraceae	*Dioscorea collettii var. hypoglauca, D. zingiberensis*	Steroidal saponins, methyl protoneogracillin, and gracillin	Hu and Yao (2003) and Tong et al. (2012)
Erythroxylaceae	*Erythroxylum pervillei*	Tropane alkaloid (pervilleine)	Mi et al. (2001, 2002, 2003)

Continued

TABLE 5.1 Plants Having Potential for Anticancer Activity—cont'd

Family	Name of Plants	Bioactive Constituent	References
Euphorbiaceae	*Croton celtidifolius, C. lechleri, C. tonkinensis, Euphorbia tirucalli, E. antiquorum, E. kansui, E, neriifolia, E. peplus, E. petiolata, E. hebecarpa, E. osyridea, E. microciadia, E. heteradenia, Excoecaria agallocha, Manihot esculenta*	Flavonols, myricetin and quercetin, flavan-3-ols of epicatechin, and epigallocatechin	Biscaro et al. (2013), Montopoli et al. (2012), Sul et al. (2013), Lin et al. (2012), Hsieh et al. (2011), Wu et al. (1991), Bigoniya and Rana (2009), Ogbourne et al. (2004), Amirghofran et al. (2011), Betancur-Galvis et al. (2002), Patil et al. (2011), Yusuf et al. (2006), Idibie et al. (2007), and Chaturvedula et al. (2003)
Fabaceae	*Anadenanthera colubrina, Arachis hypogeal, Bauhinia forficata, Copaifera langsdorffii, C. multijuga, Glycyrrhiza glabra, Psoralea corylifolia, Senna occidentalis*	Galactose and arabinose	Moretão et al. (2004), Huang et al. (2010), Ku et al. (2005), Lim et al. (2006), Costa-Lotufo et al. (2002), Gomes et al. (2008), Lima et al. (2003), Rathi et al. (2009), Honga et al. (2009), Ryu et al. (1992), Wang et al. (2011), and Calderón et al. (2006)
Fucaceae	*Ascophyllum nodosum*	Polyanionic and fucoidan	Religa et al. (2000)
Ganodermataceae	*Ganoderma lucidum*	Triterpene, ganoderic acid T, polysaccharides	Sliva et al. (2012)
Guttiferae	*Hypericum perforatum, H. adenotrichum*	Hypericin and hyperforin	Vacek et al. (2007) and Ozmen et al. (2009)
Gyrophoraceae	*Gyrophora esculenta*	Heteropolysaccharide and β-D-glucan ((1 → 6)-beta-D-glucan)	Sone et al. (1996)
Juglandaceae	*Juglans mandshurica Maxim*	Juglone	Zhang et al. (2012) and Yao et al. (2012)

Labiatae	*Salvia prionitis*	Salvicine (diterpenoid quinine)	Zhang et al. (1999) and Meng and Ding (2007)
Lamiaceae	*Mentha piperita, Scutellaria indica, Ocimum sanctum*	Eugenol, caffeic acid, rosmaric acid, and alpha- tocopherol	Kumar et al. (2004), Bonham et al. (2005), Karthikeyan et al. (1999), and Prakash and Gupta (2000)
Lecythidaceae	*Couroupita guianensis*	Isatin (1H-indole-2,3-dione)	Premanathan et al. (2012)
Leguminosae	*Astragalus membranaceus, Vicia faba*	Genistein (4′5,7-trihydroxyisoflavone), a glycoside	Auyeung et al. (2009), Wang et al. (2002), and Ravindranath et al. (2004)
Liliaceae	*Allium sativum, Aloe arborescens, A. vera*	Allicin (diallyl thiosulphate), 1,2-dimrcaptocyclopentane, isolrucin, leucine, lysine, ascorbic acid, arginine, alinine, S-allylmercaptocysteine	Ejaz et al. (2003), Lissoni et al. (2009), Furukawa et al. (2002), Jin et al. (2005), Patel et al. (2012), El-Shemy et al. (2010), Lin et al. (2006), Pecere et al. (2000), and Tong et al. (2014)
Linaceae	*Linum usitatissimum*	Secoisolariciresinol diglucoside,	Toure and Xueming (2010)
Lygodiaceae	*Lygodium flexuosum*	pheophorbide-a, pheophorbide-a′, pyropheophorbide-a, and methyl pyropheophorbide-a	Wills and Asha (2009)
Malvaceae	*Hibiscus mutabilis*	Hexameric 150-kDa lectin	Lam and Ng (2009)
Meliaceae	*Amoora rohituka, Azadirachta indica, Dysoxylum binectariferum*	C-28 Methylester derivative of triterpenoid compound, nimbolide	Rabi et al. (2013), Gogate (1991), Mohana Kumara et al. (2012), BabyKutty et al. (2012), and Hao et al. (2014)
Mimosaceae	*Mimosa pudica L*	Norepinephrine, D-pinitol, β-sitosterol, and mimosine (alkaloid)	Hullatti et al. (2013)
Moraceae	*Maclura pomifera*	Prenylated isoflavones (osajin and pomiferin)	Son et al. (2007)

Continued

TABLE 5.1 Plants Having Potential for Anticancer Activity—cont'd

Family	Name of Plants	Bioactive Constituent	References
Myrcinaceae	*Rapanea guianensis*	Khellin, berberine, lupeol, scopolin, and rapanone	Cordero et al. (2004)
Nyctaginaceae	*Boerhavia diffusa*	Punarnavine	Manu and Kuttan (2009), Leyon et al. (2005), and Ahmed-Belkacem et al. (2007)
Nyssaceae	*Camptotheca acuminata*	Camptothecin	Wall et al. (1966)
Oleaceae	*Forsythia corea*	Coreana A, B, and C	Moon et al. (1985)
Oscillatoriaceae	*Lyngbya majuscula, Symploca* spp.	Apratoxin A	Carte (1996) and Taori et al. (2008)
Papaveraceae	*Chelidonium majus, Papaver somniferum*	Thiophosphoric acid	Staniszewski et al. (1992), Ye et al. (1998), Li et al. (2012), and Sueoka et al. (1996)
Plantaginaceae	*Plantago major*	Luteolin-7-O-beta-glucoside	Gálvez et al. (2003) and Velasco-Lezama et al. (2006)
Pleurotaceae	*Pleurotus citrinopileatus*	Lectin	Li et al. (2008)
Plumbaginaceae	*Plumbago scandens, P. zeylanica*	β-Sitosterol, β-sitosteryl-3-β-glucopyranoside, β-sitosteryl-3 β-glucopyranoside-6″-O-palmitate, lupenone, lupeol acetate, plumbagin, and trilinolein	Nguyen et al. (2004), Lin et al. (2003), and Sand et al. (2012)
Poaceae	*Zea mays, Saccharum officinarum*	Tricin	Mi et al. (2016)
Polygonaceae	*Fagopyrum esculentum, Polygonum barbatum, Rheum palmatum*	Buckwheat protein	Liu et al. (2001), Abdul Mazid et al. (2011), Shoemaker et al. (2005), and Chun-Guang et al. (2010)

Polyperaceae	*Antrodia cinnamomea, Antrodia camphorate, Coriolus versicolor*	Quercetin, hyperin, rutin, protocatechuic acid, 3,4-dihydroxybenzaldehyde, vitexin, and isovitexin	Lin and Chiang (2011), Chiang et al. (2010), Chen et al. (2012), Lin and Chiang (2011), Chiang et al. (2010), and Cui and Chisti (2003)
Primulaceae	*R. guianensis*	Khellin, berberine, lupeol, scopolin, and rapanone	Cordero et al. (2004)
Ranunculaceae	*Hydrastis canadensis, Nigella sativa*	Thymoquinone and protoberberine	Sun et al. (2009) and Gali-Muhtasib et al. (2006)
Rhizophoraceae	*Caulerpa* spp., *Bruguiera sexangula*	Caulerpenyne	Fischel et al. (1995), Parent-Massin et al. (1996), Barbier et al. (2001), and Loder, and Russell (1969)
Rhodomelaceae	*Chondria* spp.	Bromoditerpene, ent-13-epiconcinndiol, and fucosterol	Palermo et al. (1992)
Rosaceae	*Duchesnea chrysantha, Prunus dulcis*	Gallic acid, methyl caffeate, protocatechuic acid and pedunculagin and brevifolin carboxylic acid	Lee and Yang (1994), Laughton et al. (1991), and Soliman and Mazzio (1998)
Rubiaceae	*Psychotria ipecacuanha, Trailliaedoxa gracilis*	Emetine, Brefeldin A, and sanguinarine chloride	Larsson et al. (2009) and Svejda et al. (2010)
Rutaceae	*Citrus limon*	Ichangensin, deoxylimonin, and obacunone	Miller et al. (2004) and Manthey and Guthrie (2002)
Scrophulariaceae	*Picrorrhiza kurroa*	Phenolic glycoside, iridoid glycoside, cucurbitacin	Rajkumar et al. (2011)
Selaginellaceae	*Selaginella moellendorffii*	biflavone, ginkgetin, biflavones (amentoflavone 7,4′,7″,4‴-tetramethyl ether, kayaflavone, podocarpusflavone A, and amentoflavone)	Sun et al. (1997)
Simaroubaceae	*Brucea antidysenterica*	Bruceantin	Cuendet and Pezzuto (2004)

Continued

TABLE 5.1 Plants Having Potential for Anticancer Activity—cont'd

Family	Name of Plants	Bioactive Constituent	References
Solanaceae	*Capsicum frutescens, Fabiana imbricata, Solanum lycocarpum, S. nigrum*	Capsanthin, capsanthin-3′-ester, and capsanthin 3,3′-diester	Maoka et al. (2001), Jun et al. (2007), Reyes et al. (2005), Munari et al. (2014), Li et al. (2009), Lee et al. (2004), Zhou et al. (2014) and Hu et al. (2013)
Taxaceae	*Taxus brevifolia, T. canadensis, T. baccata*	Taxol, vincristine, and vinblastine	Wani et al. (1971), Schiff et al. (1979), Malik et al. (2011), and Gunawardana et al. (1992)
Theaceae	*Camellia sinensis*	Epigallocatechin-3-gallate	Katiyar et al. (1992)
Umbeliferae	*Coriandrum sativum*	Phenolic content, linalool	Jana et al. (2014)
Thymelaceae	*Wikstroemia indica*	Daphnoretin (bicoumarin)	Lu et al. (2011) and Diogo et al. (2009)
Violaceae	*Viola odorata*	Varv A, varv F, and cycloviolacin (cyclotides)	Perwaiz and Sultana (1998), Lindholm et al. (2002)
Vitaceae	*Vitis vinifera*	Resveratrol (*trans*-3,4′,5-trihydroxystilbene)	Manna et al. (2000)
Zingiberaceae	*Curcuma longa*	Curcumin	Hatcher et al. (2008) and Jiao et al. (2008)
Dictyotaceae	*Stypdium* spp.	Sepitaondiol, epitaondiol diacetate, epitaondiol monoacetate, stypotriol triacetate, 14-ketostypodiol diacetate and stypodiol	Gerwick et al. (1981)

CPPS, Chlorella pyrenoidosa polysaccharides.

TABLE 5.2 Dietary Supplements With Potential Benefits in Cancer Chemotherapy

S. No.	Component	Source
1	Ginsenosides	Ginseng
2	Curcumin	Turmeric
3	Catechins/theanine	Green tea
4	Lycopene	Tomato products
5	Insoluble fiber	Wheat bran
6	Beta-glucan	Oats
7	Linoleic acid	Cheese and meat products
8	Anthrocynadins	Fruit
9	Catechins	Tea
10	Flavonones	Citrus
11	Flavones	Fruit/vegetables
12	Lignans	Rye, flax, and vegetables
13	Isoflavones, ganistein	Soya beans and soya products
14	Quercetin	Onions, apples, berries
15	Daidzein	Soya foods
16	Tangeretin	Tangerine, peels of citrus fruits

5.4. OTHER NATURAL SOURCES OF CHEMOTHERAPEUTIC AGENTS

Beside plants, an encouraging number of chemotherapeutic agents are reported in the literature from marine and microorganism sources. Many chemotherapeutic agents have been discovered from a wide range of microorganisms. Antitumor antibiotics constitute well-known cancer chemotherapeutic agents These include members of the anthracycline, bleomycin, actinomycin, mitomycin, and aureolic acid families (Newman, 2008). On the other hand, some new chemical entities have been isolated from marine organisms that have shown anticancer activity against multiple types of tumor (Basmadjian et al., 2014; Cragg and Pezzuto, 2015). In Table 5.3, potential chemotherapeutic agents from marine and microorganism sources have been enlisted.

TABLE 5.3 Chemotherapeutic Agents From Microorganism and Marine Sources

Microorganism derived	Marine sources
Actinomycin	Citarabine
Bleomycin	Bryostatin
Daunomycin	Dolastatin
Doxorubicin	Ecteinascidin
Epirubicin	Aplidine
Idarubicin	Halicondrin B
Mitomycin C	Discodermolide
Streptozocin	Cryptophycin

5.5. CONCLUSION

Keeping in view the serious side effects of conventional therapies to treat cancer, there is an overwhelming need to develop new chemoprotective/ chemotherapeutic agents that are both effective and safe. Shortlisting of plants growing in multiple but diverse hostile environments may also be considered as a vital screening strategy for the accelerated development of new chemotherapeutic agents for cancer. By understanding the complex synergistic interaction of various anticancer metabolites from diverse natural sources (plant/marine/microorganism) new chemoprotective/chemotherapeutic agents can be developed that can specifically target cancerous cells or render them benign without causing damage to normal cells. Health care professionals can also play significant clinical role in combating dreadful disease like cancer through generating awareness regarding systematic usage of herbs and other natural resources. Accelerated development of naturally derived safer chemotherapeutic agents for cancer is an immediate and crucial need of the hour.

REFERENCES

Abdul Mazid, A., Nahar, L., Datta, B.K., Khairum Bashar, S.A.M., Sarker, S.D., 2011. Potential antitumor activity of two *Polygonum species*. Arch. Biol. Sci. Belgrade 63, 465–468.

Ahmed-Belkacem, A., Macalou, S., Borrelli, F., Capasso, R., Fattorusso, E., Taglialatela-Scafati, O., Di Pietro, A., 2007. Nonprenylated rotenoids, a new class of potent breast cancer resistance protein inhibitors. J. Med. Chem. 50, 1933–1938.

Alesiani, D., Cicconi, R., Mattei, M., Montesano, C., Bei, R., Canini, A., 2008. Cell cycle arrest and differentiation induction by 5,7-dimethoxycoumarin in melanoma cell lines. Int. J. Oncol. 32, 425–434.

Amirghofran, Z., Malek-hosseini, S., Gholmoghaddam, H., Kalalinia, F., 2011. Inhibition of tumor cells growth and stimulation of lymphocytes by Euphorbia species. Immunopharmacol. Immunotoxicol. 33, 34–42.

Auyeung, K.K., Cho, C.H., Ko, J.K., 2009. A novel anticancer effect of Astragalus saponins: transcriptional activation of NSAID-activated gene. Int. J. Cancer 5, 1082–1091.

Babykutty, S., Priya, P.S., Nandini, R.J., Kumar, M.A., Nair, M.S., Srinivas, P., Gopala, S., 2012. Nimbolide retards tumor cell migration, invasion, and angiogenesis by downregulating MMP-2/9 expression via inhibiting ERK1/2 and reducing DNA-binding activity of NFkappaB in colon cancer cells. Mol. Carcinog. 51 (6), 475–490. http://dx.doi.org/10.1002/mc.20812.

Balunas, M.J., Kinghorn, A.D., 2005. Drug discovery from medicinal plants. Life Sci. 78, 431–441.

Barbier, P., Guise, S., Huitorel, P., Amade, P., Pesando, D., Briand, C., Peyrot, V., 2001. Caulerpenyne from *Caulerpa taxifolia* has an antiproliferative activity on tumor cell line SK-N-SH and modifies the microtubule network. Life Sci. 70, 415–429.

Basmadjian, C., Zhao, Q., Bentouhami, E., Djehal, A., Nebigil, C.G., Johnson, R.A., Serova, M., de Gramont, A., Faivre, S., Raymond, E., Désaubry, L.G., 2014. Cancer wars: natural products strike back. Front. Chem. 2, 20.

Beranova, L., Pombinho, A.R., Spegarova, J., Koc, M., Klanova, M., Molinsky, J., Klener, P., Bartunek, P., Andera, L., 2013. The plant alkaloid and anti-leukemia drug homoharringtonine sensitizes resistant human colorectal carcinoma cells to TRAIL-induced apoptosis via multiple mechanisms. Apoptosis. http://dx.doi.org/10.1007/s10495-013-823-829. Epub ahead of print.

Bernard, S.A., Olayinka, O.A., 2010. Search for a novel antioxidant, anti-inflammatory/analgesic or anti-proliferative drug: cucurbitacins hold the ace. J. Med. Plants Res. 4, 2821–2826.

Betancur-Galvis, L.A., Morales, G.E., Forero, J.E., Roldan, J., 2002. Cytotoxic and antiviral activities of Colombian medicinal plant extracts of the euphorbia genus. Mem. Inst. Oswaldo Cruz 97, 541–546.

Bhattacharyya, A., Chattopadhyay, R., Mitra, S., Crowe, S.E., 2014. Oxidative stress: an essential factor in the pathogenesis of gastrointestinal mucosal diseases. Physiol. Rev. 94, 329–354.

Bibi, G., Haq, I., Ullah, N., Mannan, A., Mirza, B., 2011. Antitumor, cytotoxic and antioxidant potential of *Aster thomsonii* extracts. Afr. J. Pharm. Pharmacol. 5, 252–258.

Bigoniya, P., Rana, A.C., 2009. Radioprotective and in vitro cytotoxic sapogenin from *Euphorbia neriifolia* (Euphorbiaceae) leaf. Trop. J. Pharm. Res. 8, 521–530.

Biscaro, F., Parisotto, E.B., Zanette, V.C., Günther, T.M., Ferreira, E.A., Gris, E.F., Correia, J.F., Pich, C.T., Mattivi, F., Filho, D.W., Pedrosa, R.C., 2013. Anticancer activity of flavonol and flavan-3-ol rich extracts from *Croton celtidifolius* latex. Pharm. Biol. 51, 737–743.

Bonham, M., Posakony, J., Coleman, I., Montgomery, B., Simon, J., Nelson, P.S., 2005. Characterization of chemical constituents in *Scutellaria baicalensis* with antiandrogenic and growth-inhibitory activities toward prostate carcinoma. Clin. Cancer Res. 11, 3905–3914.

Breemen, R.B.V., Pajkovic, N., 2008. Multi targeted therapy of cancer by lycopene. Cancer Lett. 269, 339–351.

Brom, B., 2009. The antioxidant controversy and cancer. South Afr. Fam. Pract. 51 (2), 119.

Butler, M.S., 2008. Natural products to drugs: natural product-derived compounds in clinical trials. Nat. Prod. Rep. 25, 475–516.

Calderón, A.I., Vázquez, Y., Solís, P.N., Caballero-George, C., Zacchino, S., Gimenez, A., Pinzón, R., Cáceres, A., Tamayo, G., Correa, M., Gupta, M.P., 2006. Screening of Latin American plants for cytotoxic activity. Pharm. Biol. 44, 130–140.

Cao, W., Li, X.Q., Wang, X., Li, T., Chen, X., Liu, S.B., Mei, Q.B., 2010. Characterizations and anti-tumor activities of three acidic polysaccharides from *Angelica sinensis* (Oliv.). Int. J. Biol. Macromol. 46, 115–122.

Carte, B.K., 1996. Biomedical potential of marine natural products. BioScience 46, 271–286.

Carter, S.K., Livingston, R.B., 1976. Plant products in cancer chemotherapy. Cancer Treat. Rep. 60, 1141–1156.

Casanova, M.L., 2003. Inhibition of skin tumor growth and angiogenesis in vivo by activation of cannabinoid receptors. J. Clin. Invest. 111, 43–50.

Cassady, J.M., Douros, J.D., 1980. Anticancer Agents Based on Natural Product Models. Academic Press, New York.

Chang, Y.S., Seo, E.K., Gyllenhaal, C., Block, K.I., 2003. *Panax ginseng*: a role in cancer therapy? Integr. Cancer Ther. 2, 13–33.

Chaturvedula, V.S.P., Schilling, J.K., Malone, S., Wisse, J.H., Werkhoven, M.C.M., Kingston, D.G., 2003. New cytotoxic triterpene acids from aboveground parts of *Manihot esculenta* from the Suriname rainforest. Planta Med. 69, 271–274.

Chen, Y.Y., Chou, P.Y., Chien, Y.C., Wu, C.H., Wu, T.S., Sheu, M.J., 2012. Ethanol extracts of fruiting bodies of *Antrodia cinnamomea* exhibit anti-migration action in human adenocarcinoma CL1-0 cells through the MAPK and PI3K/AKT signaling pathways. Phytomedicine 19, 768–778.

Chiang, P.C., Lin, S.C., Pan, S.L., Kuo, C.H., Tsai, I.L., Kuo, M.T., Wen, W.C., Chen, P., Guh, J.H., 2010. Antroquinonol displays anticancer potential against human hepatocellular carcinoma cells: a crucial role of AMPK and mTOR pathways. Biochem. Pharmacol. 79, 162–171.

Chihara, G., Hamuro, J., Maeda, Y., Arai, Y., Fukuoka, F., 1970. Fractionation and purification of the polysaccharides with marked antitumor activity, especially lentinan, from *Lentinus edodes* (Berk.) Sing. (an edible mushroom). Cancer Res. 30, 2776–2781.

Chobotova, K., Vernallis, A.B., Majid, F.A., 2010. Bromelain's activity and potential as an anti-cancer agent: current evidence and perspectives. Cancer Lett. 290, 148–156.

Christian, M.C., Wittes, R.E., Leyland-Jones, B., McLemore, T.L., Smith, A.C., Grieshaber, C.K., Chabner, B.A., Boyd, M.R., 1989. 4-Ipomeanol: a novel investigational new drug for lung cancer. J. Natl. Cancer Inst. 81, 1133–1143.

Chun-Guang, W., Jun-Qing, Y., Bei-Zhong, L., Dan-Ting, J., Chong, W., Liang, Z., Dan, Z., Yan, W., 2010. Anti-tumor activity of emodin against human chronic myelocytic leukemia K562 cell lines in vitro and in vivo. Eur. J. Pharmacol. 627, 33–41.

Cochrane, C.B., Nair, P.K.R., Melnick, S.J., Resek, A.P., Ramachandran, C., 2008. Anticancer effects of *Annona glabra* plant extracts in human leukemia cell lines. Anticancer Res. 28, 965–972.

Cordero, C.P., Gómez-González, S., León-Acosta, C.J., Morantes-Medina, S.J., Aristizabal, F.A., 2004. Cytotoxic activity of five compounds isolated from Colombian plants. Fitoterapia 75, 225–227.

Corson, T.W., Crews, C.M., 2007. Molecular understanding and modern application of traditional medicines: triumphs and trials. Cell 130, 769–774.

Costa, P.M., Ferreira, P.M., Bolzani, V.S., Furlan, M., de Freitas Formenton Macedo Dos Santos, V.A., Corsino, J., Moraes, M.O., Costa-Lotufo, L.V., Montenegro, R.C., Pessoa, C., 2008. Antiproliferative activity of pristimerin isolated from *Maytenus ilicifolia* (Celastraceae) in human HL-60 cells. Toxicol. In Vitro 22, 854–863.

Costa-Lotufo, L.V., Cunha, G.M.A., Farias, P.A.M., Viana, G.S.B., Cunha, K.M.A., Pessoa, C., Moraes, M.O., Silveira, E.R., Gramosa, N.V., Rao, V.S.N., 2002. The cytotoxic and embryotoxic effects of kaurenoic acid, a diterpene isolated from *Copaifera langsdorffii* oleo-resin. Toxicon 40, 1231–1234.

Cragg, G., Suffness, M., 1988. Metabolism of plant-derived anti-cancer agents. Pharmacol. Ther. 37, 425–461.

Cragg, G.M., Newman, D.J., Yang, S.S., 2006. Natural product extracts of plant and marine origin having antileukemia potential. The NCI experience. J. Nat. Prod. 69, 488–498.

Cragg, G.M., Pezzuto, J.M., 2015. Natural products as a vital source for the discovery of cancer chemotherapeutic and chemopreventive agents. Med. Princ. Pract. 1–19. http://dx.doi.org/10.1159/000443404.

Cragg, G.M., Boyd, M.R., Khanna, R., Kneller, R., Mays, T.D., Mazan, K.D., Newman, D.J., Sausville, E.A., 1999. International collaboration in drug discovery and development: the NCI experience. Pure Appl. Chem. 71, 1619–1633.

Cuendet, M., Pezzuto, J.M., 2004. Antitumor activity of bruceantin: an old drug with new promise. J. Nat. Prod. 67, 269–272.

Cui, J., Chisti, Y., 2003. Polysaccharopeptides of *Coriolus versicolor*: physiological activity, uses, and production. Biotechnol. Adv. 21, 109–122.

da Rocha, A.B., Lopes, R.M., Schwartsmann, G., 2001. Natural products in anticancer therapy. Curr. Opin. Pharmacol. 1, 364–369.

Dal Lago, L., D'Hondt, V., Awada, A., 2008. Selected combination therapy with sorafenib: a review of clinical data and perspectives in advanced solid tumours. Oncologist 13, 845–858.

Damayanthi, Y., Lown, J.W., 1998. Podophyllotoxins: current status and recent developments. Curr. Med. Chem. 5, 205–252.

Davis, C.D., 2007. Nutritional interactions: credentialing of molecular targets for cancer prevention. Exp. Biol. Med. 232, 176–183.

Devi, J.R., Thangam, E.B., 2012. Mechanism of anticancer activity of sulforaphane from *Brassica oleracea* in HEp-2 human epithelial carcinoma cell line. Asian Pac. J. Cancer Prev. 13, 2095–2100.

Dhuna, V., Bains, J.S., Kamboj, S.S., Singh, J., Kamboj, S., Saxena, A.K., 2005. Purification and characterization of a lectin from *Arisaema tortuosum* Schott having in-vitro anticancer activity against human cancer cell lines. J. Biochem. Mol. Biol. 38, 526–532.

Diogo, C.V., Felix, L., Vilela, S., Burgeiro, A., Barbosa, I.A., Carvalho, M.J.M., Oliveira, P.J., Peixoto, F.P., 2009. Mitochondrial toxicity of the phyotochemicals daphnetoxin and daphnoretin-relevance for possible anti-cancer application. Toxicol. In Vitro 23, 772–779.

Dipankar, C., Murugan, S., Uma Devi, P., 2011. Review on medicinal and pharmacological properties of *Iresine Herbstii*, *Chrozophora rottleri* and *Ecbolium linneanum*. Afr. J. Tradit. Complement. Altern. Med. 8, 124–129.

Dorr, R.T., Dvorakova, K., Snead, K., Alberts, D.S., Salmon, S.E., Pettit, G.R., 1996. Antitumor activity of combretastatin-A4 phosphate, a natural product tubulin inhibitor. Invest. New Drugs 14, 131–137.

Duan, H., Luan, J., Liu, Q., Yagasaki, K., Zhang, G., 2010. Suppression of human lung cancer cell growth and migration by berbamine. Cytotechnology 62, 341–348.

Dutt, R., Garg, V., Madan, A., 2014. Can plants growing in diverse hostile environments provide a vital source of anticancer drugs? Cancer Ther. 10, 10–37.

Efferth, T., Olbrich, A., Sauerbrey, A., Ross, D.D., Gebhart, E., Neugebauer, M., 2002. Activity of ascaridol from the anthelmintic herb *Chenopodium anthelminticum* L. against sensitive and multidrug-resistant tumor cells. Anticancer Res. 22, 4221–4224.

Ejaz, S., Woong, L.C., Ejaz, A., 2003. Extract of garlic (*Allium sativum*) in cancer chemoprevention. Exp. Oncol. 25, 93–97.

El-Sayed, A., Cordell, G.A., 1981. Catharanthus alkaloids. XXXIV. Catharanthamine, a new antitumor bisindole alkaloid from *Catharanthus roseus*. J. Nat. Prod. 44, 289–293.

El-Sayed, A., Handy, G.A., Cordell, G.A., 1983. Catharanthus alkaloids, XXXVIII. Confirming structural evidence and antineoplastic activity of the bisindole alkaloids leurosine-N'b-oxide (pleurosine), roseadine and vindolicine from *Catharanthus roseus*. J. Nat. Prod. 46, 517–527.

El-Sayed, A.M., Ezzat, S.M., Salama, M.M., Sleem, A.A., 2011. Hepatoprotective and cytotoxic activities of *Delonix regia* flower extracts. Pharmacogn. J. 3, 49–56. http://dx.doi.org/10.5530/pj.2011.19.10.

El-Shemy, H.A., Aboul-Soud, M.A., Nassr-Allah, A.A., Aboul-Enein, K.M., Kabash, A., Yagi, A., 2010. Antitumor properties and modulation of antioxidant enzyme's activity by *Aloe vera* leaf active principles isolated via supercritical carbon dioxide extraction. Curr. Med. Chem. 17, 129–138.

Endringer, D.C., Pezzuto, J.M., Braga, F.C., 2009. NF-κB inhibitory activity of cyclitols isolated from *Hancornia speciosa*. Phytomed 16, 1064–1069.

Epifano, F., Genovese, S., Fiorito, S., Mathieu, V., Kiss, R., 2014. Lapachol and its congeners as anticancer agents: a review. Phtyochem. Rev. 13, 37–49.

Ferreira, P.M.P., Farias, D.F., Viana, M.P., Souza, T.M., Vasconcelos, I.M., Soares, B.M., Pessoa, C., Costa-Lotufo, L.V., Moraes, M.O., Carvalho, A.F.U., 2011. Study of the antiproliferative potential of seed extracts from Northeastern Brazilian plants. An. Acad. Bras. Ciênc. 83, 1045–1058.

Fischel, J.L., Lemee, R., Formento, P., Caldani, C., Moll, J.L., Pesando, D., Meinesz, A., Grelier, P., Pietra, P., Guerriero, A., 1995. Cell growth inhibitory effects of caulerpenyne, a sesquiterpenoid from the marine algae *Caulerpa taxifolia*. Anticancer Res. 15, 2155–2160.

Forgo, P., Zupkó, I., Molnár, J., Vasas, A., Dombi, G., Hohmann, J., 2012. Bioactivity-guided isolation of antiproliferative compounds from *Centaurea jacea* L. Fitoterapia 83, 921–925.

Fulda, S., 2008. Betulinic Acid for cancer treatment and prevention. Int. J. Mol. Sci. 9, 1096–1107.

Furukawa, F., Nishikawa, A., Chihara, T., Shimpo, K., Beppu, H., Kuzuya, H., Lee, I.S., Hirose, M., 2002. Chemopreventive effects of *Aloe arborescens* on N-nitrosobis (2-oxopropyl) amine-induced pancreatic carcinogenesis in hamsters. Cancer Lett. 178, 117–122.

Gali-Muhtasib, H., Roessner, A., Schneider-Stock, R., 2006. Thymoquinone: a promising anti-cancer drug from natural sources. Int. J. Biochem. Cell Biol. 38, 1249–1253.

Gálvez, M., Martín-Cordero, C., López-Lázaro, M., Cortés, F., Ayuso, M.J., 2003. Cytotoxic effect of *Plantago* spp. on cancer cell lines. J. Ethnopharmacol. 88, 125–130.

Gerwick, W.H., Fenical, W., 1981. Ichthyotoxic and cytotoxic metabolites of the tropical brown alga *Stypopodium zonale* (Lamouroux) papenfuss. J. Org. Chem. 46, 22–27.

Gogate, S.S., 1991. Cytotoxicity of neem leaf extract: an antitumor. Natl. Med. J. India 9, 297.

Gomes, N.M., Rezende, C.M., Fontes, S.P., Hovell, A.M.C., Landgraf, R.G., Matheus, M.E., Pinto, A.C., Fernandes, P.D., 2008. Antineoplasic activity of *Copaifera multijuga* oil and fractions against ascitic and solid Ehrlich tumor. J. Ethnopharmacol. 119, 179–184.

Gunawardana, G.P., Premachandran, U., Burres, N.S., Whittern, D.N., Henry, R., Spanton, S., McAlpine, J.B., 1992. Isolation of 9-dihydro-13-acetylbaccatin III from *Taxus canadensis*. J. Nat. Prod. 55, 1686–1689.

Habtemariam, S., Macpherson, A.M., 2000. Cytotoxicity and antibacterial activity of ethanol extract from leaves of a herbal drug, boneset (*Eupatorium perfoliatum*). Phytother. Res. 14, 575–577.

Hao, F., Kumar, S., Yadav, N., Chandra, D., 2014. Neem components as potential agents for cancer prevention and treatment. Biochim. Biophys. Acta 1846 (1), 247–257. http://dx.doi.org/10.1016/j.bbcan. 2014. 07.002.

Harvey, A.L., 2008. Natural products in drug discovery. Drug Discov. Today 13, 894–901.

Hatcher, H., Planalp, R., Cho, J., Torti, F.M., Torti, S.V., 2008. Curcumin: from ancient medicine to current clinical trials. Cell. Mol. Life Sci. 65, 1631–1652.

Hoessel, R., Leclerc, S., Endicott, J.A., Nobel, M.E.M., Lawrie, A., Tunnah, P., Leost, M., Damiens, E., Marie, D., Marko, D., Niederberger, E., Tang, W., Eisenbrand, G., Meijer, L.,

1999. Indirubin, the active constituent of a Chinese antileukaemia medicine, inhibits cyclin-dependent kinases. Nat. Cell Biol. 1, 60–67.

Honga, Y.K., Wub, H.T., Mab, T., Liuc, W.J., He, X.J., 2009. Effects of *Glycyrrhiza glabra* polysaccharides on immune and antioxidant activities in high-fat mice. Int. J. Biol. Macromol. 45, 61–64.

Hsieh, W.T., Lin, H.Y., Chen, J.H., Kuo, Y.H., Fan, M.J., Wu, R.S., Wu, K.C., Wood, W.G., Chung, J.G., 2011. Latex of *Euphorbia antiquorum* induces apoptosis in human cervical cancer cells via c-jun n-terminal kinase activation and reactive oxygen species production. Nutr. Cancer 63, 1339–1347.

Hu, B., An, H.M., Shen, K.P., Shi, X.F., Deng, S., Wei, M.M., 2013. Effect of *Solanum nigrum* on human colon carcinoma RKO cells. Zhong Yao Cai 36, 958–961. PMID:24380285.

Hu, K., Yao, X., 2003. The cytotoxicity of methyl protoneogracillin (NSC-698793) and gracillin (NSC-698787), two steroidal saponins from the rhizomes of *Dioscorea collettii* var. *hypoglauca*, against human cancer cells in vitro. Phytother Res. 17, 620–626.

Huang, C.P., Au, L.C., Chiou, R.Y., Chung, P.C., Chen, S.Y., Tang, W.C., Chang, C.L., Fang, W.H., Lin, S.B., 2010. Arachidin-1, a peanut stilbenoid, induces programmed cell death in human leukemia HL-60 cells. J. Agric. Food Chem. 58, 12123–12129.

Huang, Y.G., Li, Q.Z., Ivanochko, G., Wang, R., 2006. Novel selective cytotoxicity of wild sarsaparilla rhizome extract. J. Pharm. Pharmacol. 58, 1399–1403.

Hullatti, K., Pathade, N., Mandavkar, Y., Godavarthi, A., Biradi, M., 2013. Bioactivity-guided isolation of cytotoxic constituents from three medicinal plants. Pharm. Biol. 51, 601–606.

Huntimer, E.D., Halaweish, F.T., Chase, C.C., 2006. Proliferative activity of *Echinacea angustifolia* root extracts on cancer cells: interference with doxorubicin cytotoxicity. Chem. Biodivers. 3, 695–703.

Hussain, H., Krohn, K., Ahmad, V.U., Miana, G.A., Green, I.R., 2007. Lapachol: an overview. ARKIVOC 2, 145–171.

Idibie, C.A., Davids, H., Iyuke, S.E., 2007. Cytotoxicity of purified cassava linamarin to a selected cancer cell lines. Bioprocess Biosyst. Eng. 30, 261–269.

Jana, S., Patra, K., Sarkar, S., Jana, J., Mukherjee, G., Bhattacharjee, S., Mandal, D.P., 2014. Antitumorigenic potential of linalool is accompanied by modulation of oxidative stress: an in vivo study in sarcoma-180 solid tumor model. Nutr. Cancer 66, 835–848. http://dx.doi.org/10.1080/01635581.2014.904906.

Jayaprakasam, B., Seeram, N.P., Nair, M.G., 2003. Anticancer and antiinflammatory activities of cucurbitacins from *Cucurbita andreana*. Cancer Lett. 189, 11–16.

Jiao, Y., Wilkinson, J., Di, X., Wang, W., Hatcher, H., Kock, N.D., D'Agostino Jr., R., Knovich, M.A., Torti, F.M., Torti, S.V., 2008. Curcumin, a cancer chemopreventive and chemotherapeutic agent, is a biologically active iron chelator. Blood 113, 462–469.

Jin, G.Z., Quan, H.J., Koyanagi, J., Takeuchi, K., Miura, Y., Komada, F., Saito, S., 2005. 4'-O-Alkyaloenin derivatives and their sulfates directed toward overcoming multidrug resistance in tumor cells. Cancer Lett. 218, 15–20.

Jun, H.S., Park, T., Lee, C.K., Kang, M.K., Park, M.S., Kang, H., Surh, Y.J., Kim, O.H., 2007. Capsaicin induced apoptosis of B16-F10 melanoma cells through down-regulation of Bcl-2. Food Chem. Toxicol. 45, 708–715.

Karakas, F.P., Yildirim, A., Turker, A., 2012. Biological screening of various medicinal plant extracts for antibacterial and antitumor activities. Turk. J. Biol. 36, 641–652.

Karthikeyan, K., Gunasekaran, P., Ramamurthy, N., Govindasamy, S., 1999. Anticancer activity of *Ocimum sanctum*. Pharm. Biol. 37, 285–290.

Katiyar, C., Gupta, A., Kanjilal, S., Katiyar, S., 2012. Drug discovery from plant sources: an integrated approach. Ayu 33, 10–19.

Katiyar, S.K., Agarwal, R., Wang, Z.Y., Bhatia, A.K., Mukhtar, H., 1992. (–)-Epigallocatechin-3-gallate in *Camellia sinensis* leaves from Himalayan region of Sikkim: inhibitory effects against biochemical events and tumor initiation in Sencar mouse skin. Nutr. Cancer 18, 73–83.

Kerbel, R., Folkman, J., 2002. Clinical translation of angiogenesis inhibitors. Nat. Rev. Cancer 2, 727–739.

Ku, K.L., Chang, P.S., Cheng, Y.C., Lien, C.Y., 2005. Production of stilbenoids from the callus of *Arachis hypogaea*: a novel source of the anticancer compound piceatannol. J. Agric. Food Chem. 53, 3877–3881.

Kubo, I., Ochi, M., Vieira, P.C., Komatsu, S., 1993. Antitumor agents from the cashew (*Anacardium occidentale*) apple juice. J. Agric. Food Chem. 41, 1012–1015.

Kumar, A., Samarth, R.M., Yasmeen, S., Sharma, A., Sugahara, T., Terado, T., Kimura, H., 2004. Anticancer and radioprotective potentials of *Mentha piperita*. Biofactors 22, 87–91.

Kuo, P.L., Hsu, Y.L., Chang, C.H., Lin, C.C., 2005. The mechanism of ellipticine-induced apoptosis and cell cycle arrest in human breast MCF-7 cancer cells. Cancer Lett. 223, 293–301.

Lam, S.K., Ng, T.B., 2009. Novel galacturonic acid-binding hexameric lectin from *Hibiscus mutabilis* seeds with antiproliferative and potent HIV-1 reverse transcriptase inhibitory activities. Acta Biochim. Pol. 56, 649–654.

Larsson, D.E., Hassan, S., Larsson, R., Oberg, K., Granberg, D., 2009. Combination analyses of anticancer drugs on human neuroendocrine tumor cell lines. Cancer Chemother. Pharmacol. 65, 5–12.

Laughton, M.J., Evans, P.J., Moroney, M.A., Hoult, J.R., Halliwell, B., 1991. Inhibition of mammalian 5-lipoxygenase and cyclooxygenase by flavonoids and phenolic dietary additives. Relationship to antioxidant activity and to iron ionreducing ability. Biochem. Pharmacol. 42, 1673–1681.

Lee, I.R., Yang, M.Y., 1994. Phenolic compounds from *Duchesnea chrysantha* and their cytotoxic activities in human cancer cell. Arch. Pharmcal. Res. 17, 476–479.

Lee, I.S., Shamon, L.A., Chai, H.B., Chagwedera, T.E., Besterman, J.M., Farnsworth, N.R., Cordell, G.A., Pezzuto, J.M., Kinghorn, A.D., 1996. Cell-cycle specific cytotoxicity mediated by rearranged *ent*-kaurene diterpenoids isolated from *Parinari curatellifolia*. Chem. Biol. Interact. 99, 193–204.

Lee, K.H., Huang, H.C., Huang, E.S., Furukawa, H., 1972. Antitumor agents II: Eupatolide, a new cytotoxic principle from *Eupatorium formosanum* hay. J. Pharm. Sci. 61, 629–631.

Lee, S.J., Oh, P.S., Ko, J.H., Lim, K., Lim, K.T., 2004. A 150- kDa glycoprotein isolated from *Solanum nigrum* L. has cytotoxic and apoptotic effects by inhibiting the effects of protein kinase C alpha, nuclear factor-kappa B and inducible nitric oxide in HCT-116 cells. Cancer Chemother. Pharmacol. 54, 562–572.

Leyon, P.V., Lini, C.C., Kuttan, G., 2005. Inhibitory effect of *Boerhaavia diffusa* on experimental metastasis by B16F10 melanoma in C57BL/6 mice. Life Sci. 76, 1339–1349.

Li, J., Li, Q.W., Gao, D.W., Han, Z.S., Lu, W.Z., 2009. Antitumor and immunomodulating effects of polysaccharides isolated from *Solanum nigrum* Linne. Phytother. Res. 23, 1524–1530.

Li, S., He, J., Li, S., Cao, G., Tang, S., Tong, Q., Joshi, H.C., 2012. Noscapine induced apoptosis via down regulation of survivin in human neuroblastoma cells having wild type or null p53. PLoS One 7, e40076.

Li, Y.R., Liu, Q.H., Wang, H.X., Ng, T.B., 2008. A novel lectin with potent antitumor, mitogenic and HIV-1 reverse transcriptase inhibitory activities from the edible mushroom *Pleurotus citrinopileatus*. Biochim. Biophys. Acta 1780, 51–57.

Lim, H., Kim, M.K., Lim, Y., Cho, Y.H., Lee, C.H., 2006. Inhibition of cell-cycle progression in HeLa cells by HY52, a novel cyclin-dependent kinase inhibitor isolated from *Bauhinia forficata*. Cancer Lett. 233, 89–97.

Lima, S.R., Junior, V.F., Christo, H.B., Pinto, A.C., Fernandes, P.D., 2003. In vivo and in vitro studies on the anticancer activity of *Copaifera multijuga* hayne and its fractions. Phytother. Res. 17, 1048–1053.

Lin, J.G., Chen, G.W., Li, T.M., Chouh, S.T., Tan, T.W., Chung, J.G., 2006. Aloe-emodin induces apoptosis in T24 human bladder cancer cells through the p53 dependent apoptotic pathway. J. Urol. 175, 343–347.

Lin, L.C., Yang, L.L., Chou, C.J., 2003. Cytotoxic naphthoquinones and plumbagic acid glucosides from *Plumbago zeylanica*. Phytochemistry 62, 619–622.

Lin, M.W., Lin, A.S., Wu, D.C., Wang, S.S., Chang, F.R., Wu, Y.C., Huang, Y.B., 2012. Euphol from *Euphorbia tirucalli* selectively inhibits human gastric cancer cell growth through the induction of ERK1/2-mediated apoptosis. Food Chem. Toxicol. 50, 4333–4339.

Lin, Y.W., Chiang, B.H., 2011. 4-acetylantroquinonol B isolated from *Antrodia cinnamomea* arrests proliferation of human hepatocellular carcinoma HepG2 cell by affecting p53, p21 and p27 levels. J. Agric. Food Chem. 59, 8625–8631.

Lindholm, P., Gullbo, J., Claeson, P., Göransson, U., Johansson, S., Backlund, A., Larsson, R., Bohlin, L., 2002. Selective cytotoxicity evaluation in anticancer drug screening of fractionated plant extracts. J. Biomol. Screen. 7, 333–340.

Lissoni, P., Rovelli, F., Brivio, F., Zago, R., Colciago, M., Messina, G., Mora, A., Porro, G., 2009. A randomized study of chemotherapy versus biochemotherapy with chemotherapy plus *Aloe arborescens* in patients with metastatic cancer. In vivo 23, 171–175.

Liu, Z., Ishikawa, W., Huang, X., Tomotake, H., Kayashita, J., Watanabe, H., Kato, N., 2001. A buckwheat protein product suppresses 1,2-Dimetylhydrazine-induced colon carcinogenesis in rats by reducing cell proliferation. J. Nutr. 131, 1850–1853.

Loder, J.W., Russell, G.B., 1969. Tumor inhibitory plants. The alkaloids of *Bruguiera sexangula* and *Bruguiera exaristata* (Rhizophoraceae). Aus. J. Chem. 22, 1271–1275.

Lone, S.H., Bhat, K.A., Khuroo, M.A., 2015. Arglabin: from isolation to antitumor evaluation. Chem. Biol. Interact. 240, 180–198.

Lu, C.L., Li, Y.M., Fu, G.Q., Yang, L., Jiang, J.G., Zhu, L., Lin, F.L., Chen, J., Lin, Q.S., 2011. Extraction optimisation of daphnoretin from root bark of *Wikstroemia indica* (L.) C.A. and its antitumour activity tests. Food Chem. 124, 1500–1506.

Ma, X., Wang, Z., 2009. Anticancer drug discovery in the future: an evolutionary perspective. Drug Discov. Today 14, 1137–1142.

Machado, F.B., Yamamoto, R.E., Zanoli, K., Nocchi, S.R., Novello, C.R., Schuquel, I.T.A., Sakuragui, C.M., Luftmann, H., Ueda-Nakamura, T., Nakamura, C.V., de Mello, J.C., 2012. Evaluation of the antiproliferative activity of the leaves from *Arctium lappa* by a bioassay-guided fractionation. Molecules 17, 1852–1859.

Malik, S., Cusidó, R.M., Mirjalili, M.H., Moyano, E., Palazón, J., Bonfill, M., 2011. Production of the anticancer drug taxol in *Taxus baccata* suspension cultures: a review. Process Biochem. 46, 23–34.

Manna, S.K., Mukhopadhyay, A., Aggarwal, B.B., 2000. Resveratrol suppresses TNF-induced activation of nuclear transcription factors NF-kappa B, activator protein-1, and apoptosis: potential role of reactive oxygen intermediates and lipid peroxidation. J. Immunol. 164, 6509–6519.

Manthey, J.A., Guthrie, N., 2002. Antiproliferative activities of citrus flavonoids against six human cancer cell lines. J. Agric. Food Chem. 21, 5837–5843.

Manu, K.A., Kuttan, G., 2009. Anti-metastatic potential of Punarnavine, an alkaloid from *Boerhaavia diffusa* Linn. Immunobiology 214, 245–255.

Maoka, T., Mochida, K., Kozuka, M., Ito, Y., Fujiwara, Y., Hashimoto, K., Enjo, F., Ogata, M., Nobukuni, Y., Tokuda, H., Nishino, H., 2001. Cancer chemopreventive activity of carotenoids in the fruits of red paprika *Capsicum annuum* L. Cancer Lett. 172, 103–109.

Meng, L.H., Ding, J., 2007. Salvicine, a novel topoisomerase II inhibitor, exerts its potent anticancer activity by ROS generation. Acta Pharmacol. Sin. 28, 1460–1465.

Mi, L., Pu, Y., Chang Geun, Y., Ragauskas, A.J., 2016. The occurrence of Tricin and its derivatives in plants. Green Chem. 18, 1439–1454.

Mi, Q., Cui, B., Lantvit, D., Reyes-Lim, E., Chai, H., Pezzuto, J.M., Kinghorn, A.D., Swanson, S.M., 2003. Pervilleine, F a new tropane alkaloid aromatic ester that reverses multidrug resistance. Anticancer. Res. 23, 3607–3615.

Mi, Q., Cui, B., Silva, G.L., Lantvit, D., Lim, E., Chai, H., Hollingshead, M.G., Mayo, J.G., Kinghorn, A.D., Pezzuto, J.M., 2002. Pervilleines B and C, new tropane alkaloid aromatic esters that reverse the multidrug-resistance in the hollow fiber assay. Cancer. Lett. 184, 13–20.

Mi, Q., Cui, B., Silva, G.L., Lantvit, D., Lim, E., Chai, H., You, M., Hollingshead, M.G., Mayo, J.G., Kinghorn, A.D., Pezzuto, J.M., 2001. Pervilleine A, a novel tropane alkaloid that reverses the multidrug resistance phenotype. Cancer Res. 61, 4030–4037.

Miller, E.G., Porter, J.L., Binnie, W.H., Guo, I.Y., Hasegawa, S., 2004. Further studies on the anticancer activity of *Citrus limonoids*. J. Agric. Food Chem. 52, 4908–4912.

Mohana Kumara, P., Zuehlke, S., Priti, V., Ramesha, B.T., Shweta, S., Ravikanth, G., Vasudeva, R., Santhoshkumar, T.R., Spiteller, M., Uma Shaanker, R., 2012. *Fusarium proliferatum*, an endophytic fungus from *Dysoxylum binectariferum* Hook.f, produces rohitukine, a chromane alkaloid possessing anti-cancer activity. Antonie Van Leeuwenhoek 101, 323–329.

Montopoli, M., Bertin, R., Chen, Z., Bolcato, J., Caparrotta, L., Froldi, G., 2012. Croton lechleri sap and isolated alkaloid taspine exhibit inhibition against human melanoma SK23 and colon cancer HT29 cell lines. J. Ethnopharmacol. 144, 747–753.

Moon, C.K., Park, K.S., Lee, S.H., Ha, B.J., Gon-Lee, B., 1985. Effects of antitumor polysaccharides from *Forsythia Corea* on the immune function (I). Arch. Pharm. Res. 8, 31–38.

Moretão, M.P., Zampronio, A.R., Gorin, P.A., Iacomini, M., Oliveira, M.B., 2004. Induction of secretory and tumoricidal activities in peritoneal macrophages activated by an acidic heteropolysaccharide (ARAGAL) from the gum of *Anadenanthera colubrina* (*Angico branco*). Immunol. Lett. 93, 189–197.

Munari, C.C., de Oliveira, P.F., Campos, J.C., Martins, S.D., Da Costa, J.C., Bastos, J.K., Tavares, D.C., 2014. Antiproliferative activity of *Solanum lycocarpum* alkaloidic extract and their constituents, solamargine and solasonine, in tumor cell lines. J. Nat. Med. 68, 236–241.

Munson, A.E., Harris, L.S., Friedman, M.A., Dewey, W.L., Carchman, R.A., 1975. Antineoplastic activity of cannabinoids. J. Natl. Cancer Inst. 55, 597–602.

Nascimento, F.R., Cruz, G.V., Pereira, P.V., Maciel, M.C., Silva, L.A., Azevedo, A.P., Barroqueiro, E.S., Guerra, R.N., 2006. Ascitic and solid Ehrlich tumor inhibition by *Chenopodium ambrosioides* L. treatment. Life Sci. 78, 2650–2653.

Newman, D.J., Cragg, G.M., 2007. Natural products as sources of new drugs over the last 25 years. J. Nat. Prod. 70, 461–477.

Newman, D.J., 2008. Natural products as leads to potential drugs: an old process or the new hope for drug discovery? J. Med. Chem. 51, 2589–2599.

Nguyen, A.T., Malonne, H., Duez, P., Vanhaelen-Fastre, R., Vanhaelen, M., Fontaine, J., 2004. Cytotoxic constituents from *Plumbago zeylanica*. Fitoterapia 75, 500–504.

Ogbourne, S.M., Suhrbier, A., Jones, B., Cozzi, S.J., Boyle, G.M., Morris, M., McAlpine, D., Johns, J., Scott, T.M., Sutherland, K.P., Gardner, J.M., Le, T.T., Lenarczyk, A., Aylward, J.H., Parsons, P.G., 2004. Antitumor activity of 3-ingenyl angelate: plasma membrane and mitochondrial disruption and necrotic cell death. Cancer Res. 64, 2833–2839.

Ohnuma, T., Holland, J.F., 1985. Homoharringtonine as a new antileukemic agent. J. Clin. Oncol. 3, 604–606.

Ooi, K.L., Muhammad, T.S., Tan, M.L., Sulaiman, S.F., 2011. Cytotoxic, apoptotic and anti-α-glucosidase activities of 3,4-di-O-caffeoyl quinic acid, an antioxidant isolated from the polyphenolic-rich extract of *Elephantopus mollis* Kunth. J. Ethnopharmacol. 135, 685–695.

Ordóñez, P.E., Sharma, K.K., Bystrom, L.M., Alas, M.A., Enriquez, R.G., Malagón, O., Jones, D.E., Guzman, M.L., Compadre, C.M., 2016. Dehydroleucodine, a sesquiterpene Lactone from *Gynoxys verrucosa*, demonstrates cytotoxic activity against human leukemia cells. J. Nat. Prod. 79 (4), 691–696.

Osiecki, H., 2002. Cancer: A Nutritional, Biochemical Approach. Bioconcepts Publishing.

Ozmen, A., Bauer, S., Gridling, M., Singhuber, J., Krasteva, S., Madlener, S., Vo, T.P., Stark, N., Saiko, P., Fritzer-Szekeres, M., Szekeres, T., Askin-Celik, T., Krenn, L., Krupitza, G., 2009. In vitro anti-neoplastic activity of the ethno-pharmaceutical plant *Hypericum adenotrichum* Spach endemic to Western Turkey. Oncol. Rep. 4, 845–852.

Palermo, J.A., Flower, P.B., Seldes, A.M., 1992. Chondriamides A and B, new indolic metabolites from the red alga Chondria sp. Tetrahedron Lett. 33, 3097–3100.

Pan, L., Chai, H., Kinghorn, A.D., 2010. The continuing search for antitumor agents from higher plants. Phytochem. Lett. 3, 1–8.

Pan, S.Y., Zhou, S.F., Gao, S.H., Yu, Z.L., Zhang, S.F., Tang, M.K., Sun, J.N., Ma, D.L., Han, Y.F., Fong, W.F., Ko, K.M., 2013. New perspectives on how to discover drugs from herbal medicines: CAM's outstanding contribution to modern therapeutics. Evid. Based Complement. Altern. Med. 2013, 627375.

Paoletti, C., Le Pecq, J.B., Dat-Xuong, N., Juret, P., Garnier, H., Amiel, J.L., Rouesse, J., 1980. Antitumor activity, pharmacology, and toxicity of ellipticines, ellipticinium and 9-hydroxy-derivatives: preliminary clinical trials of 2-methyl-9- hydroxy-ellipticinium (NSC 264-137). Recent Res. Cancer Res. 74, 107–123.

Parada-Turska, J., Paduch, R., Majdan, M., Kandefer-Szerszeñ, M., Rzeski, W., 2007. Antiproliferative activity of parthenolide against three human cancer cell lines and human umbilical vein endothelial cells. Pharmacol. Rep. 59, 233–237.

Parent-Massin, D., Fournier, V., Amade, P., Lemée, R., Durand-Clément, M., Delescluse, C., Pesando, D., 1996. Evaluation of the toxicological risk to humans of caulerpenyne using human hematopoietic progenitors, melanocytes, and keratinocytes in culture. J. Toxicol. Environ. Health 47, 47–59.

Park, H.B., Lee, K.H., Kim, K.H., Lee, K., Noh, H.J., Choi, S.U., Lee, K.R., 2009. Lignans from the roots of *Berberis amurensis*. Nat. Prod. Sci. 15, 17–21.

Patel, D.K., Patel, K., Tahilyani, V., 2012. Barbaloin: a concise report of its pharmacological and analytical aspects. Asian Pac. J. Trop. Biomed. 2, 835–838.

Patil, R.C., Manohar, S.M., Upadhye, M.V., Katchi, V.I., Rao, A.J., Mule, A., Moghe, A.S., 2011. Anti reverse transcriptase and anticancer activity of stem ethanol extracts of *Excoecaria agallocha* (Euphorbiaceae). Ceylon J. Sci. (Biol. Sci.) 40, 147–155.

Patočka, J., 2003. Biologically active pentacyclic triterpenes and their current medicine signification. J. Appl. Biomed. 1, 7–12.

Pecere, T., Gazzola, M.V., Mucignat, C., Parolin, C., Vecchia, F.D., Cavaggioni, A., Basso, G., Diaspro, A., Salvato, B., Carli, M., Palú, G., 2000. Aloe-emodin is a new type of anticancer agent with selective activity against neuroectodermal tumors. Cancer Res. 60, 2800–2804.

Persinos, G.P., Blomster, R.N., 1978. South American plants III: isolation of fulvoplumierin from *Himatanthus sucuuba* (M. Arg.) Woodson (Apocynaceae). J. Pharm. Sci. 67, 1322–1323.

Perwaiz, S., Sultana, S., 1998. Antitumorigenic effect of crude extract of *Viola odorata* on DMBA-induced two stage skin carcinogenesis in the Swiss albino mice. Asia Pac. J. Pharmacol. 13, 43–50.

Philips, B.U., 1999. The case for Cancer nutritional support. Cancer Nutr. Netw. Tex.

Powell, R.G., Weisleder, D., Smith Jr., C.R., Rohwedder, W.K., 1970. Structures of harringtonine, isoharringtonine, and homoharringtonine. Tetrahedron Lett. 11, 815–818.

Prakash, J., Gupta, S.K., 2000. Chemopreventive activity of *Ocimum sanctum* seed oil. J. Ethnopharmacol. 72, 29–34.

Premanathan, M., Radhakrishnan, S., Kulangiappar, K., Singaravelu, G., Thirumalaiarasu, V., Sivakumar, T., Kathiresan, K., 2012. Antioxidant and anticancer activities of isatin (1H-indole-2,3-dione), isolated from the flowers of *Couroupita guianensis* Aubl. Indian J. Med. Res. 136, 822–826.

Rabi, T., Huwiler, A., Zangemeister-Wittke, U., 2013. AMR-Me inhibits PI3K/Akt signaling in hormone-dependent MCF-7 breast cancer cells and inactivates NF-κB in hormone-independent MDA-MB-231 cells. Mol. Carcinog.. http://dx.doi.org/10.1002/mc.22012. Epub ahead of print.

Rajendran, N.N., Deepa, N., 2007. Anti-tumor activity of *Acanthospermum hispidum* DC on dalton ascites lymphoma in mice. Nat. Prod. Sci. 13, 234–240.

Rajkumar, V., Guha, G., Kumar, R.A., 2011. Antioxidant and anti-neoplastic activities of *Picrorhiza kurroa* extracts. Food Chem. Toxicol. 49, 363–369.

Rao, K., McBride, T.J., Oleson, J.J., 1968. Recognition and evaluation of lapachol as an antitumor agent. Cancer Res. 28, 1952–1954.

Rathi, S.G., Suthar, M., Patel, P., Bhaskar, V.H., Rajgor, N.B., 2009. In-vitro cytotoxic screening of *Glycyrrhiza glabra* L. (Fabaceae): a natural anticancer drug. J. Young Pharm. 1, 239–243.

Ravindranath, M.H., Muthugounder, S., Presser, N., Viswanathan, S., 2004. Anticancer therapeutic potential of soy isoflavone, genistein. Adv. Exp. Med. Biol. 546, 121–165.

Rebecca, S., Deepa, N., Ahmedin, J., 2012. Cancer statistics, 2012. CA Cancer J. Clin. 62, 10–29.

Rebouças Sde, O., Grivicich, I., Dos Santos, M.S., Rodriguez, P., Gomes, M.D., de Oliveira, S.Q., da Silva, J., Ferraz Ade, B., 2011. Antiproliferative effect of a traditional remedy, *Himatanthus articulatus* bark, on human cancer cell lines. J. Ethnopharmacol. 137, 926–929.

Religa, P., Kazi, M., Thyberg, J., Gaciong, Z., Swedenborg, J., Hedin, U., 2000. Fucoidan inhibits smooth muscle cell proliferation and reduces mitogen-activated protein kinase activity. Eur. J. Vasc. Endovasc. Surg. 20, 419–426.

Rennó, M.N., Barbosa, G.M., Zancan, P., Veiga, V.F., Alviano, C.S., Sola-Penna, M., Menezes, F.S., Holandino, C., 2008. Crude ethanol extract from babassu (*Orbignya speciosa*): cytotoxicity on tumoral and non-tumoral cell lines. An. Acad. Bras. Ciênc. 80, 467–476.

Reyes, M., Schmeda-Hirschmann, G., Razmilic, I., Theoduloz, C., Yáñez, T., Rodríguez, J.A., 2005. Gastroprotective activity of sesquiterpene derivatives from *Fabiana imbricate*. Phytother. Res. 19, 1038–1042.

Rivera, J.O., Loya, A.M., Ceballos, R., 2013. Use of herbal medicines and implications for conventional drug therapy medical sciences. Altern. Integ. Med. 2, 130.

Robinson, M.M., Zhang, X., 2011. Traditional Medicines: Global Situation, Issues and Challenges. The World Medicines Situation, third ed. WHO, Geneva, pp. 1–14.

Roja, G., Rao, P.S., 2000. Anticancer compounds from tissue cultures of medicinal plants. J. Herbs Spices Med. Plants 7, 71–102.

Roman, G.P., Neagu, E., Moroeanu, V., Radu, G.L., 2008. Concentration of *Symphytum officinale* extracts with cytostatic activity by tangential flow ultra filtration. Roum. Biotechnol. Lett 13, 4008–4013.

Ryu, S.Y., Choi, S.U., Lee, C.O., Zee, O.P., 1992. Antitumor activity of *Psoralea corylifolia*. Arch. Pharma. Res. 15, 356–359.

Sagar, S.M., Yance, D., Wong, R.K., 2006. Natural health products that inhibit angiogenesis: a potential source for investigational new agents to treat cancer part 2. Curr. Oncol. 13, 99–107.

Sak, K., 2012. Chemotherapy and dietary phytochemical agents. Chemother. Res. Pract. 282570, 11.

Saklani, A., Kutty, S.K., 2008. Plant-derived compounds in clinical trials. Drug Discov. Today 13, 161–171.

Sand, J.M., Bin Hafeez, B., Jamal, M.S., Witkowsky, O., Siebers, E.M., Fischer, J., Verma, A.K., 2012. Plumbagin (5-hydroxy-2-methyl-1,4-naphthoquinone), isolated from *Plumbago zeylanica*, inhibits ultraviolet radiation-induced development of squamous cell carcinomas. Carcinogenesis 33, 184–190.

Sashidhara, K.V., White, K.N., Crews, P., 2009. A selective account of effective paradigms and significant outcomes in the discovery of inspirational marine natural products. J. Nat. Prod. 72, 588–603.

Schiff, P.B., Fant, J., Horwitz, S.B., 1979. Promotion of microtubule assembly in vitro by taxol. Nature 277, 665–667.

Seo, J.M., Kang, H.M., Son, K.H., Kim, J.H., Lee, C.W., Kim, H.M., Chang, S.I., Kwon, B.M., 2003. Antitumor activity of flavones isolated from *Artemisia argyi*. Planta Med. 69, 218–222.

Shan, B.E., Yoshita, Y., Sugiura, T., Yamashita, U., 1999. Suppressive effect of Chinese medicinal herb, *Acanthopanax gracilistylus* extract on human lymphocytes in vitro. Clin. Exp. Immunol. 118, 41–48.

Sheng, J., Yu, F., Xin, Z., Zhao, L., Zhu, X., Hu, Q., 2007. Preparation, identification and their antitumor activities in vitro of polysaccharides from *Chlorella pyrenoidos*. Food Chem. 105, 533–539.

Shirota, O., Morita, H., Takeya, K., Itokawa, H., 1994. Cytotoxic aromatic triterpenes from *Maytenus ilicifolia* and *Maytenus chuchuhuasca*. J. Nat. Prod. 57, 1675–1681.

Shoeb, M., 2006. Anticancer agents from medicinal plants. Bangladesh. J. Pharmacol. 1, 35–41.

Shoemaker, M., Hamilton, B., Dairkee, S.H., Cohen, I., Campbell, M.J., 2005. In vitro anticancer activity of twelve Chinese medicinal herbs. Phytother. Res. 19, 649–651.

Sliva, D., Loganathan, J., Jiang, J., Jedinak, A., Lamb, J.G., Terry, C., Baldridge, L.A., Adamec, J., Sandusky, G.E., Dudhgaonkar, S., 2012. Mushroom *Ganoderma lucidum* prevents colitis-associated carcinogenesis in mice. PLoS One 7, e47873. http://dx.doi.org/10.1371/journal.pone.0047873.

Soliman, K.F., Mazzio, E.A., 1998. In vitro attenuation of nitric oxide production in C6 astrocyte cell culture by various dietary compounds. Proc. Soc. Exp. Biol. Med. 218, 390–397.

Son, H.I., Chung, I.M., Lee, S.I., Yang, H.D., Moon, H.-I., 2007. Pomiferin, histone deacetylase inhibitor isolated from the fruits of *Maclura pomifera*. Biorg. Med. Chem. Lett. 17, 4753–4755.

Sone, Y., Isoda-Johmura, M., Misaki, A., 1996. Isolation and chemical characterization of polysaccharides from Iwatake, *Gyrophora esculenta* Miyoshi. Biosci. Biotechnol. Biochem. 60, 213–215.

Stähelin, H., 1973. Activity of a new glycosidic lignan derivative (VP-16-213) related to podophyllotoxin in experimental tumors. Eur. J. Cancer 9, 215–221.

Staniszewski, A., Slesak, B., Kołodziej, J., Harłozińska-Szmyrka, A., Nowicky, J.W., 1992. Lymphocyte subsets in patients with lung cancer treated with thiophosphoric acid alkaloid derivatives from *Chelidonium majus* L. (Ukrain). Drugs Exp. Clin. Res. 18, 63–67.

Sueoka, N., Sueoka, E., Okabe, S., Fujiki, H., 1996. Anti-cancer effects of morphine through inhibition of tumour necrosis factor-a release and mRNA expression. Carcinogenesis 17, 2337−2341.

Sul, Y.H., Lee, M.S., Cha, E.Y., Thuong, P.T., Khoi, N.M., Song, I.S., 2013. An ent-kaurane diterpenoid from Croton tonkinensis induces apoptosis by regulating AMP-activated protein kinase in SK-HEP1 human hepatocellular carcinoma cells. Biol. Pharm. Bull. 36, 158−164.

Sun, C.M., Syu, W.J., Huang, Y.T., Chen, C.C., Ou, J.C., 1997. Selective cytotoxicity of ginkgetin from *Selaginella moellendorffii*. J. Nat. Prod. 60, 382−384.

Sun, Y., Xun, K., Wang, Y., Chen, X., 2009. A systematic review of the anticancer properties of berberine, a natural product from chinese herbs. Anticancer Drugs 20, 757−769.

Svejda, B., Aguiriano-Moser, V., Sturm, S., Hoger, H., Ingolic, E., Siegel, V., Stuppner, H., Pfragner, R., 2010. Anticancer activity of novel plant extracts from *Trailliaedoxa gracilis* (W.W. Smith and Forrest) in human carcinoid KRJ-1 cells. Anticancer Res. 30, 55−64.

Svoboda, G.H., Poore, G.A., Montfort, M.L., 1968. Alkaloids of *Ochrosia maculata* Jacq. (*Ochrosia borbonica* Gmel.). Isolation of the alkaloids and study of the antitumor properties of 9-methoxyellipticine. J. Pharm. Sci. 57, 1720−1725.

Talib, W.H., 2011. Anticancer and antimicrobial potential of plant-derived natural products. In: Rasooli, I. (Ed.), Phytochemicals Bioactivities and Impact on Health. InTech, pp. 141−158. http://dx.doi.org/10.5772/26077.

Taori, K., Paul, V.J., Luesch, H., 2008. Structure and activity of largazole, a potent antiproliferative agent from the Floridian marine *Cyanobacterium symploca* sp. J. Am. Chem. Soc. 130, 1806−1807.

Tímár, J., Csuka, O., Orosz, Z., Jeney, A., Kopper, L., 2001. Molecular pathology of tumor metastasis. I. Predictive pathology. Pathol. Oncol. Res. 7, 217−230.

Tong, D., Qu, H., Meng, X., Jiang, Y., Liu, D., Ye, S., Chen, H., Jin, Y., Fu, S., Geng, J., 2014. S-allylmercaptocysteine promotes MAPK inhibitor-induced apoptosis by activating the TGF-β signaling pathway in cancer cells. Oncol. Rep. 32, 1124−1132. http://dx.doi.org/10.3892/or.2014.3295.

Tong, Q.Y., He, Y., Zhao, Q.B., Qing, Y., Huang, W., Wu, X.H., 2012. Cytotoxicity and apoptosis-inducing effect of steroidal saponins from *Dioscorea zingiberensis* Wright against cancer cells. Steroids 77, 1219−1227.

Toure, A., Xueming, X., 2010. Flaxseed lignans: source, biosynthesis, metabolism, antioxidant activity, bio-active components, and health benefits. Compr. Rev. Food Sci. Food Saf. 9, 261−269.

Tyagi, A., Bhatia, N., Condon, M.S., Bosland, M.C., Agarwal, C., Agarwal, R., 2002. Antiproliferative and apoptotic effects of silibinin in rat prostate cancer cells. Prostate 53, 211−217.

Ukiya, M., Akihisa, T., Yasukawa, K., Tokuda, H., Suzuki, T., Kimura, Y., 2006. Anti-inflammatory, anti-tumor-promoting, and cytotoxic activities of constituents of marigold (*Calendula officinalis*) flowers. J. Nat. Prod. 69, 1692−1696.

Vacek, J., Klejdus, B., Kubán, V., 2007. Hypericin and hyperforin: bioactive components of St. John's Wort (*Hypericum perforatum*). Their isolation, analysis and study of physiological effect. Ceska. Slov. Farm. 56, 62−66.

Velasco-Lezama, R., Tapia-Aguilar, R., Román-Ramos, R., Vega-Avila, E., Pérez-Gutiérrez, M.S., 2006. Effect of *Plantago major* on cell proliferation in vitro. J. Ethnopharmacol. 103, 36−42.

Wall, M.E., Wani,, M.C., 1996. Camptothecin and taxol: from discovery to clinic. J. Ethnopharmacol. 51, 239−253.

Wall, M.E., Wani, M.C., Cook, C.E., Palmer, K.H., Mcphail, A.T., Sim, G.A., 1966. Plant antitumor agents: I. The isolation and structure of camptothecin, a novel alkaloidal leukemia and tumor inhibitor from *Camptotheca acuminata*. J. Am. Chem. Soc. 88, 3888−3890.

Wang, H.Z., Zhang, Y., Xie, L.P., Yu, X.Y., Zhang, R.Q., 2002. Effects of genistein and daidzein on the cell growth, cell cycle, and differentiation of human and murine melanoma cells. J. Nutr. Biochem. 13, 421–426.

Wang, J., Li, Q., Ivanochko, G., Huang, Y., 2006. Anticancer effect of extracts from a North American medicinal plant-wild Sarsaparilla. Anticancer Res. 26, 2157–2164.

Wang, Y., Hong, C., Zhou, C., Xu, D., Qu, H.B., 2011. Screening antitumor compounds Psoralen and Isopsoralen from *Psoralea corylifolia* L. Seeds. Evid. Based Complement. Altern. Med. 2011, 363052. http://dx.doi.org/10.1093/ecam/nen087.

Wani, M.C., Taylor, H.L., Wall, M.E., Coggon, P., McPhail, A.T., 1971. Plant antitumor agents. VI. The isolation and structure of taxol, a novel antileukemic and antitumor agent from *Taxus brevifolia*. J. Am. Chem. Soc. 93, 2325–2327.

Wills, P.J., Asha, V.V., 2009. Chemopreventive action of *Lygodium flexuosum* extract in human hepatoma PLC/PRF/5 and Hep 3B cells. J. Ethnopharmacol. 122, 294–303.

Wong, K.F., Yuan, Y., Luk, J.M., 2012. *Tripterygium wilfordii* bioactive compounds as anticancer and anti-inflammatory agents. Clin. Exp. Pharmacol. Physiol. 39, 311–320.

Wu, T.S., Lin, Y.M., Haruna, M., Pan, D.J., Shingu, T., Chen, Y.P., Hsu, H.Y., Nakano, T., Lee, K.H., 1991. Antitumor Agents 119. kansuiphorins A and B, two novel antileukemic diterpene esters from *Euphorbia kansui*. J. Nat. Prod. 54, 823–829.

Xie, J., Ma, T., Gu, Y., Zhang, X., Qiu, X., Zhang, L., Xu, R., Yu, Y., 2009. Berbamine derivatives: a novel class of compounds for anti-leukemia activity. Eur. J. Med. Chem. 44, 3293–3298.

Xie, S.S., 1989. Immunoregulatory effect of polysaccharide *of Acanthopanax senticosus* (PAS). 1. Immunological mechanism of PAS against cancer. Zhonghua Zhong Liu Za Zhi 11, 338–340.

Xu, R., Dong, Q., Yu, Y., Zhao, X., Gan, X., Wu, D., Lu, Q., Xu, X., Yu, X.F., 2006a. Berbamine: a novel inhibitor of bcr/abl fusion gene with potent anti-leukemia activity. Leuk. Res. 30, 17–23.

Xu, Y., Smith, J.A., Lannigan, D.A., Hecht, S.M., 2006b. Three acetylated flavonol glycosides from *Forsteronia refracta* that specifically inhibit p90 RSK. Bioorg. Med. Chem. 14, 3974–3977.

Yao, Y., Zhang, Y.W., Sun, L.G., Liu, B., Bao, Y.L., Lin, H., Zhang, Y., Zheng, L.H., Sun, Y., Yu, C.L., Wu, Y., Wang, G.N., Li, Y.X., 2012. Juglanthraquinone C, a novel natural compound derived from *Juglans mandshurica* Maxim, induces S phase arrest and apoptosis in HepG2 cells. Apoptosis 17, 832–841.

Ye, K., Ke, Y., Keshava, N., Shanks, J., Kapp, J.A., Tekmal, R.R., Petros, J., Joshi, H.C., 1998. Opium alkaloid noscapine is an antitumor agent that arrests metaphase and induces apoptosis in dividing cells. Proc. Natl. Acad. Sci. USA 95, 1601–1606.

Yusuf, U.F., Ahmadun, F.R., Rosli, R., Iyuke, S.E., Billa, N., Abdullah, N., Umar-Tsafe, N., 2006. An in vitro inhibition of human malignant cell growth of crude water extract of cassava (*Manihot esculenta* Crantz) and commercial linamarin. Nutraceutical. Funct. Food 28, 145–155.

Zhang, H., Pei, Z., Zhang, X., Kang, T., 2010. Antitumor activities of liposoluble components of caulis *Marsdeniae tenocissimae* and analysis on its chemical constituents. Zhongguo Zhong Yao Za Zhi 35, 3325–3328.

Zhang, J.S., Ding, J., Tang, Q.M., Li, M., Zhao, M., Lu, L.J., Chen, L.J., Yuan, S.T., 1999. Synthesis and antitumor activity of novel diterpene quinone salvicine and the analogs. Bioorg. Med. Chem. Lett. 9, 2731–2736.

Zhang, W., Liu, A., Li, Y., Zhao, X., Lv, S., Zhu, W., Jin, Y., 2012. Anticancer activity and mechanism of juglone on human cervical carcinoma HeLa cells. Can. J. Physiol. Pharmacol. 90, 1553–1558.

Zheng, G.Q., Kenney, P.M., Lam, L.K., 1992. Anethofuran, carvone and limonene: potential cancer chemoprotective agents from dill weed oil and caraway oil. Planta Med. 58, 338–341.

Zhou, Y., Tang, Q., Zhao, S., Zhang, F., Li, L., Wu, W., Wang, Z., Hann, S., 2014. Targeting signal transducer and activator of transcription 3 contributes to the solamargine-inhibited growth and -induced apoptosis of human lung cancer cells. Tumour Biol. 35, 8169–8178. PMID:24845028.

Zhu, H.H., Huang, G.H., Tate, P.L., Larcom, L.L., 2012. Anti cancer effect of *Angelica sinensis* on women's reproductive cancer. Funct. Foods Health Dis. 2, 242–250.

Zhu, W.J., Yu, D.H., Zhao, M., Lin, M.G., Lu, Q., Wang, Q.W., Guan, Y.Y., Li, G.X., Luan, X., Yang, Y.F., Qin, X.M., Fang, C., Yang, G.H., Chen, H.Z., 2013. Antiangiogenic triterpenes isolated from Chinese herbal medicine *Actinidia chinensis* Planch. Anticancer Agents Med. Chem. 13, 195–198.

Chapter 6

Speeding Up the Virtual Design and Screening of Therapeutic Peptides: Simultaneous Prediction of Anticancer Activity and Cytotoxicity

A. Speck-Planche, M.N.D.S. Cordeiro
University of Porto, Porto, Portugal

6.1. INTRODUCTION

Cancers are life-threatening medical conditions characterized by the uncontrolled growth of genetically damaged cells, which can spread to different parts of the body. Cancers are one of the leading causes of mortality and morbidity all over the world. In fact, according to the GLOBOCAN project, about 14.1 million new cancer cases and 8.2 million deaths occurred in 2012 (Torre et al., 2015). Meanwhile in 2014, 14.1 million people were estimated to develop cancer annually (McGuire, 2016). Over time, the burden has been concentrated in less developed countries, which currently account for approximately 57% of cases and 65% of cancer deaths worldwide. Lung cancer has been the leading cause of cancer death among males in both more and less developed countries, and it has surpassed breast cancer as the leading cause of cancer death among females in more developed countries (Torre et al., 2015); breast cancer has remained the leading cause of cancer death among females in less developed countries. In addition, other frequently diagnosed cancers with high incidence include those of the liver, stomach, and colorectum among males, while cancers of the stomach, cervix, uteri, and colorectum have been the most prevalent among females (Torre et al., 2015).

In the past years, the pharmaceutical industry has focused its efforts on providing efficacious anticancer agents by the process known as drug discovery. However, the design and development of any new drug is a very expensive, complex, and time-consuming task, which can take around 13.5 years, with an

Multi-Scale Approaches in Drug Discovery. http://dx.doi.org/10.1016/B978-0-08-101129-4.00006-0

estimated cost of $1.8 billion for each new molecular entity that comes to the market (Paul et al., 2010). On the other hand, the multifactorial nature of cancers (Martinez-Lacaci et al., 2007), and the emergence of phenomena such as multidrug resistance, make this group of diseases very difficult to treat. Additionally, anticancer drugs are well known for their serious adverse effects (Cheng and Force, 2010; Illouz et al., 2014; Torino et al., 2013; Verges et al., 2014).

All the aforementioned ideas have paved the way for the rationalization of the drug discovery process in cancer research. Within this context, peptides have been considered as promising therapeutic alternatives due to their potent and broad spectrum of biological activities (de la Fuente-Nunez et al., 2014; Hein et al., 2015; Nguyen et al., 2011; Zhao et al., 2016), including inhibitory potency against many types of cancer cell lines (Chu et al., 2015; Karpel-Massler et al., 2016). Despite their multiple applications, peptides can be rapidly degraded by proteolytic environments (Gorris et al., 2009), and therefore, stability is an aspect of great importance, which should be considered when designing new peptides. Nevertheless, the stability of peptides can be enhanced through different ways (Chen et al., 2012; Gentilucci et al., 2010), such as changing the backbone chemistry, incorporating α-aminoxy amino acids and/or D-amino acids, cyclization, or by introducing protecting groups at the termini of the peptides. Anyway, these chemical modifications may not be enough to reduce the cytotoxicity, which remains one of the main concerns in the early development of peptide-based drugs (Maher and McClean, 2006, 2008).

The accelerated growth in the potential use of peptides as emerging pharmacological agents has led to the creation of web repositories devoted to the study of the structures, functions, and activity profiles of the peptides (Agrawal et al., 2016; Gautam et al., 2014; Gogoladze et al., 2014; Kapoor et al., 2012; Kumar et al., 2015; Mehta et al., 2014; Pirtskhalava et al., 2016; Tyagi et al., 2015). Noteworthy advances have been achieved in the field of chemoinformatics and bioinformatics through the establishment of diverse computational models focused on the prediction of anticancer peptides (Chen et al., 2016; Dennison et al., 2010; E-Kobon et al., 2016; Hajisharifi et al., 2014; Tyagi et al., 2013), while only a few *in silico* approaches have been applied to the identification of toxic and nontoxic peptides (Chaudhary et al., 2016; Gupta et al., 2013).

At least one of the following handicaps is found in the current models used for the discovery of peptide-based therapeutics in cancer research. First, most of the models reported to date predict anticancer activity unspecifically, that is, without considering the different cancer cell lines against which the peptides are active. Therefore, the probabilities of discovering potent anticancer peptides may be remarkably reduced. The same may occur if in the predictions, only one cancer cell line is targeted. Second, the physicochemical variables (known as molecular descriptors) computed from the three-dimensional structures or the sequences of the peptides have a global nature, which prevents the analysis of the regions responsible for the enhancement of the anticancer activity. Similar drawbacks are found in computational models devoted to the prediction of toxic and nontoxic peptides, where the toxicity is predicted without specifying the cell against which a defined peptide might or might not be toxic. In an attempt to overcome all these disadvantages, several authors have suggested the use of multitasking (mtk) chemoinformatic models, a class of advanced *in silico* tools that are able to perform simultaneous predictions of multiple biological effects (activities, toxicities, etc.) against diverse living systems (microorganisms and cells, among others) (Romero-Duran et al., 2016; Speck-Planche and Cordeiro, 2014; Tenorio-Borroto et al., 2014). In the present chapter, we describe a novel computational methodology aimed at setting up an mtk-chemoinformatic model for the virtual design and screening of peptides with potential anticancer activity against different cancer cell lines, and low cytotoxicity against diverse healthy mammalian cells.

6.2. MATERIALS AND METHODS

6.2.1. Dataset and Calculation of the Molecular Descriptors

All chemical and biological data regarding the anticancer activities of the peptides were retrieved from the web repository known as CancerPPD (Tyagi et al., 2015), which is a database containing a compilation of experimental assays focused on the anticancer activities of peptides and proteins. At the same time, all experimental data focused on cytotoxicity assays were extracted from the public source named Database of Antimicrobial Activity and Structure of Peptides (Gogoladze et al., 2014).

The data set used in this study is composed of 1108 peptides, containing from 2 to 69 amino acids. Each peptide was assayed against at least 1 of 35 biological targets (b_t), which included cancer cell lines and healthy mammalian cells. In each assay, at least one of three measures of biological effect (m_e) was employed, namely: the inhibitory concentration at 50% (IC_{50}), cytotoxic concentration at 50% (CC_{50}), and hemolytic concentration at 50% (HC_{50}). In the data set, all the IC_{50}, CC_{50}, and HC_{50} were expressed in micromolar, by this means allowing for an accurate comparison of the anticancer activities or cytotoxicities among the peptides. Not all the peptides were reported against all the biological targets. Consequently, the data set used here contained only 1933 cases of peptides, i.e., combinations involving the peptide plus the measure of its biological effect and the corresponding biological target. Each peptide case in the data set was then assigned to one of two possible categories/classes, namely, positive [$BE_q(c_r) = 1$, indicating potent anticancer activity or low cytotoxicity] or negative [$BE_q(c_r) = -1$, referred to low anticancer activity or high cytotoxicity]. The cutoff values to annotate a peptide as positive were: $IC_{50} \leq 25$ μM, $CC_{50} \geq 80.37$ μM, or $HC_{50} \geq 105.7$ μM.

Here, $BE_q(c_r)$ is a binary variable characterizing the biological effect of the qth peptide under the experimental condition c_r. Notice that the experimental condition c_r is an ontology of the form $c_r \rightarrow (m_e, b_t)$, where m_e and b_t have been previously explained.

All the peptide sequences were stored in a file of type *.txt, the latter then being transformed to *.fasta file through the use of a format converter, that is, by the online tool belonging to the web repository named HIV Sequence Database (Gaschen et al., 2001). Thereafter, the software ProtDCal was employed to calculate the Broto−Moreau autocorrelations [$ACk(w)$] (Ruiz-Blanco et al., 2015), a family of molecular descriptors that has been widely studied in the scientific literature (Todeschini and Consonni, 2000). The configuration used to calculate the Broto−Moreau autocorrelations is summarized in the supplementary material (file **SM1.pdf**), which can be accessed via the companion website associated with this book. The $ACk(w)$ indices can be calculated according to the following form:

$$ACk(w) = \sum_{i=1}^{N}\sum_{j=1}^{N} w_i \cdot w_j \cdot \delta(d_{ij}; k) \tag{6.1}$$

In Eq. (6.1), N is the total number of amino acids and w_i and w_j are the physicochemical properties of the amino acids i and j, respectively. In addition, δ is the Kronecker delta; $\delta = 1$ if $d_{ij} = k$ and $L \geq k$, where d_{ij} is the topological distance between the ith and jth amino acids, k is the cutoff value of topological distance, and L is the length of the peptide (number of amino acids). In the case where $d_{ij} < k$ and $L > 1$, the $ACk(w)$ index is calculated as a constant value, being the simple sum of the squares of the physicochemical properties of each amino acid. In this study, from the $ACk(w)$ indices, new molecular descriptors were obtained:

$$MACk(w) = \frac{ACk(w) \cdot \theta^{nt} \cdot K^{ct}}{L} \tag{6.2}$$

In Eq. (6.2), $MACk(w)$ represents the modified Broto−Moreau autocorrelations, while nt and ct are binary variables characterizing the presence or absence of chemical modifications at the N-terminus (acetyl) and the C-terminus (amino), respectively. The term L is employed to normalize the $ACk(w)$ indices. In Eq. (6.2), the arbitrary mathematical terms $\theta = 1.306$ (Mills's constant) and $K = 1.132$ (Viswanath's constant) have been introduced with the aim of increasing/decreasing the values of the $MACk(w)$ indices according to the chemical modifications present in the N- and C-termini, respectively. Notice that when $\theta > K$, it means that the acetyl group modifying the N-terminus is larger than the amino group modifying the C-terminus. Thus, with the use of θ and K, there are no drastic changes in the $MACk(w)$ values, but the values change enough to differentiate unmodified peptides from those with modifications at the termini. Here, we would like to point out that other mathematical constants may also be employed.

6.2.2. Box–Jenkins Approach and the Creation of the Multitasking Chemoinformatics Model

A close look at Eq. (6.2) discloses that the $MACk(w)$ indices only take into account the peptide sequences. Therefore, these molecular descriptors will not be able to differentiate the anticancer activity or the cytotoxicity of a peptide when assayed against diverse cancer cell lines or different healthy mammalian cells, respectively. Nonetheless, the Box–Jenkins approach that encompasses the calculation of moving averages offers a simple solution to overcome such a problem (Hill and Lewicki, 2006). Nowadays, Box–Jenkins operators have been applied to several works focused on drug discovery (Romero-Duran et al., 2016; Speck-Planche et al., 2016). In the first step regarding the calculation of the Box–Jenkins moving averages, the following expression is used:

$$\text{avg\_}MACk(w)c_r = \frac{1}{n(c_r)} \sum_{q=1}^{n(c_r)} MACk_q(w) \qquad (6.3)$$

In Eq. (6.3), $n(c_r)$ is the number of peptides tested by considering the same element of the experimental condition (ontology) c_r, which have also been annotated as positive. For example, in the case of the element b_t, $n(c_r)$ will be the number of peptides tested against the same biological target, which have been assigned as positive. The same deduction can be made for the element m_e. Also, in Eq. (6.3), $MACk_q(w)$ is the modified autocorrelation index for the qth peptide, while $avg\_MACk(w)c_r$ is the arithmetic mean of the $MACk_q(w)$ indices. After calculating the $avg\_MACk(w)c_r$ values, a subsequent expression can be used for each peptide:

$$DMACk_q(w)c_r = MACk_q(w) - \text{avg\_}MACk(w)c_r \qquad (6.4)$$

Here, in Eq. (6.4), the deviation term $DMACk_q(w)$ is an adaptation of the Box–Jenkins operators, and it considers both the peptide sequence and a defined element of the experimental condition c_r under which the peptide was assayed. For this reason, only the $DMACk_q(w)$ indices (210 in total) were taken into account for setting up the mtk-chemoinformatic model.

The data set formed by the 1933 cases of peptides was randomly partitioned into two series, i.e., training and prediction (test) sets. The training set was employed to find the best model. This set was formed by 1469 cases, 676 assigned as positive and 793 as negative. On the other hand, the prediction set contained 464 cases, 211 annotated as positive and the remaining 253 assigned as negative. This set was employed to assess the predictive power of the mtk-chemoinformatic model. Artificial neural networks (ANN) was used as the data analysis method for generating the model, and diverse ANN architectures were explored, namely, linear neural networks, multilayer perceptron, radial basis function (RBF), and probabilistic neural networks. Correlations between variables were analyzed. Thus, a cutoff interval $-0.7 < r < 0.7$ was applied as a criterion for lack of redundancy, r being the Pearson's correlation coefficient (Pearson, 1895). All data analysis was performed by the program STATISTICA v6.0 (Statsoft-Team, 2001), including a sensitivity analysis intended to select the best variables (molecular descriptors). In so doing, the neural network module of STATISTICA applied a missing value substitution procedure, which was used to allow predictions to be made in the absence of values for one or more input variables (Statsoft-Team, 2001). Thus, to define the sensitivity of a particular input variable (molecular descriptor) v, each ANN was run on the training set cases, and a network error was generated (Hill and Lewicki, 2006). Thereafter, the network was run again using the same cases, but this time replacing the observed values of v with the value estimated by the missing value procedure. Consequently, a new network error was generated. Thus, the sensitivity of each molecular descriptor was calculated as the ratio of the error with missing value substitution to the original error. The more sensitive the network was to a particular molecular descriptor, the greater the ratio. It should be pointed out here that only descriptors with high sensitivity values (>1) were selected, and we ensured that at least one variable belonging to each element of the ontology c_r was among the chosen variables.

The performance of the model was determined through the analysis of several statistical indices such as percentages of correct classification for positive and negative cases, accuracy, the Matthews's correlation coefficient (MCC), and the areas under the receiver operating characteristic (ROC) curves (Romero-Duran et al., 2016; Speck-Planche et al., 2016). The general steps employed for the generation of the mtk-chemoinformatic model are depicted in Fig. 6.1.

6.3. RESULTS AND DISCUSSION

6.3.1. mtk-Chemoinformatic Model Based on Artificial Neural Networks

We took into account the fact that the best model should be the one exhibiting a very high statistical quality but containing as few descriptors as possible. The best mtk-chemoinformatic model based on ANN had the following profile: RBF 6:6-200-1:1. This indicates that the model is based on an ANN with an RBF architecture, using six molecular descriptors from which the input layer was generated (also containing six nodes or input vectors), with a subsequent hidden layer (200 neurons) containing the activation functions, and one output layer, which serves to give the predicted value of the biological effect [Pred-$BE_q(c_r)$]. The symbols and definitions of the different molecular descriptors that entered in the final model are shown in Table 6.1.

The model correctly classified 639 of 676 (94.53%) positive cases of peptides in the training set. In the same set, 740 of 793 (93.32%) cases of negative peptides were properly classified. Here, the accuracy was 93.87%. At the same

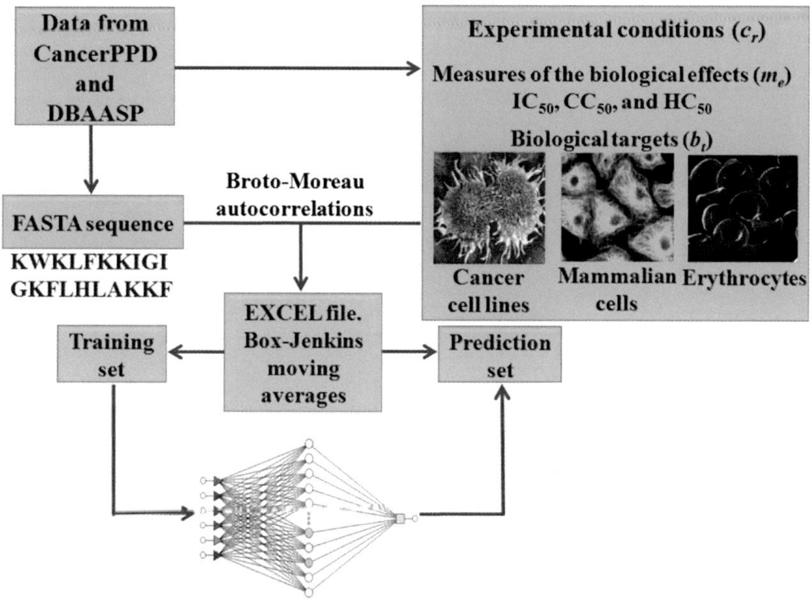

FIGURE 6.1 General steps describing the creation of the multitasking chemoinformatic model.

time, 196 of 211 positive (92.89%) and 235 of 253 negative (92.89%) cases of peptides were correctly classified in the prediction (test) set, thus yielding an accuracy of 92.89%. Data focused on the input file to develop the mtk-chemoinformatic model (**SM2.xlsx**) and the ANN file (**SM3.snn**) are provided in the supplementary materials. At the same time, all chemical and biological data, as well as the particular statistics for each peptide, are included in the file **SM4.xlsx**. All these supplementary materials can be accessed via the companion website associated with this book. We should stress here that all these percentages of correct classification give only a global idea regarding the performance of the model. Therefore, we have also calculated local statistical indices pertaining to the percentages of correctly classified peptides, depending on the different biological targets [%CCP(b_t)] against which the peptides were tested, (see Table 6.2). The %CCP(b_t) values ranged in the intervals 75−100% for positive cases, and 78.57−100% for negative cases, respectively, whereas all the %CCP(m_e) values were higher than 93% and 79.8% for positive and negative cases, respectively.

In addition to the statistical indices mentioned earlier, MCC was also used as another measure of the great performance of the model. The MCC values were 0.877 and 0.857 for the training and prediction sets, respectively. These values are close to 1, which means that there is a strong correlation between

TABLE 6.1 Molecular Descriptors Present in the Final Multitasking Chemoinformatic Model

Symbol	Definition
$DMAC1(Mw)m_e$	Deviation of the Broto–Moreau autocorrelation weighted by the molar mass of each amino acid in unfolded state, considering all the amino acids placed at topological distance equal to 1, and depending on the measures of the biological effects
$DMAC2(Anp)m_e$	Deviation of the Broto–Moreau autocorrelation weighted by the nonpolar area of each amino acid in unfolded state, considering all the amino acids placed at topological distance equal to 2, and depending on the measures of the biological effects
$DMLAC5(Z1)m_e$	Deviation of the Broto–Moreau autocorrelation weighted by the combined measure of hydrophobicity-related properties of each amino acid in unfolded state, considering all the amino acids placed at topological distance equal to 5, and depending on the measures of the biological effects
$DMAC2(Mw)b_t$	Deviation of the Broto–Moreau autocorrelation weighted by the molar mass of each amino acid in unfolded state, considering all the amino acids placed at topological distance equal to 2, and depending on the biological targets against which a peptide was assayed
$DMAC6(Ap)b_t$	Deviation of the Broto–Moreau autocorrelation weighted by the polar area of each amino acid in unfolded state, considering all the amino acids placed at topological distance equal to 6, and depending on the biological targets against which a peptide was assayed
$DMAC7(Anp)b_t$	Deviation of the Broto–Moreau autocorrelation weighted by the nonpolar area of each amino acid in unfolded state, considering all the amino acids placed at topological distance equal to 7, and depending on the biological targets against which a peptide was assayed

the observed and predicted values of the categorical variable $BE_q(c_r)$. As a final proof of the statistical quality and predictive power of the model, we calculated the areas under the ROC curves (Fig. 6.2). The values of areas under the ROC curves were 0.985 for the training set and 0.977 for the prediction set. Notice that these values demonstrate that the present mtk-chemoinformatic model does not behave as a random classifier, for which the areas under the ROC curves are equal to 0.5. Altogether, the analysis of the diverse statistical indices confirms the high quality and predictive capability of the model.

TABLE 6.2 Group Statistics Associated With the Multitasking Chemoinformatic Model for the Diverse Biological Targets (b_t)

Biological Target (b_t)	NP_{Total} (b_t)	$NP_{Positive}$ (b_t)	$NP_{Negative}$ (b_t)	$\%CCP_{Positive}$ (b_t)	$\%CCP_{Negative}$ (b_t)
Lung cancer cells (NCI-H128)	8	7	1	100	100
Prostate cancer cells (PC-3)	128	23	105	86.96	96.19
Lung cancer cells (NCI-H69)	8	7	1	100	100
Lung cancer cells (NCI-H146)	8	7	1	100	100
Cervical cancer cells (HeLa)	131	21	110	95.24	93.64
Colorectal cancer cells (SW-480)	110	16	94	87.5	98.94
Lung cancer cells (H-1299)	122	21	101	85.71	95.05
Breast cancer cells (MDA-MB-361)	19	8	11	100	100
Leukemia cancer cells (K-562)	23	15	8	93.33	87.5
Breast cancer cells (MCF-7)	139	31	108	96.77	96.30
Colorectal cancer cells (SW-1116)	9	5	4	100	100
Skin cancer cells (BMKC)	110	9	101	100	96.04
Renal cancer cells (A498)	6	6	0	100	–
Leukemia cancer cells (Jurkat)	19	10	9	100	100
Skin cancer cells (A375)	9	5	4	100	100
Skin cancer cells (B-16)	9	5	4	100	100
Lung cancer cells (A549)	26	14	12	100	100

Continued

TABLE 6.2 Group Statistics Associated With the Multitasking Chemoinformatic Model for the Diverse Biological Targets (b_t)—cont'd

Biological Target (b_t)	NP_{Total} (b_t)	$NP_{Positive}$ (b_t)	$NP_{Negative}$ (b_t)	$\%CCP_{Positive}$ (b_t)	$\%CCP_{Negative}$ (b_t)
Rhabdomyosarcoma cancer cells (RD)	8	4	4	100	100
Colorectal cancer cells (HCT-116)	6	6	0	100	–
Leukemia cancer cells (CCRF-CEM)	9	6	3	83.33	100
Mouse fibroblast cells (NIH 3T3)	21	10	11	100	100
MARC-145cells	11	11	0	90.91	–
Murine macrophage cells (raw 264.7)	6	4	2	75	100
Madin–Darby canine kidney cells	5	0	5	–	100
Murine fibroblast cells (L929)	9	3	6	100	83.33
Murine splenocytes	9	0	9	–	100
Human monocytic THP-1cells	33	28	5	100	100
Human umbilical vein endothelial cells	35	6	29	83.33	100
Rat epithelial cells (IEC-6, CRL-1592)	26	7	19	100	100
CEM-SS cells	61	0	61	–	100
Sheep erythrocytes	20	9	11	100	100
Rabbit erythrocytes	18	18	0	83.33	–
Human erythrocytes	663	498	165	93.57	78.79
Rat erythrocytes	99	57	42	98.25	78.57
Mouse erythrocytes	10	10	0	90	–

%CCP, percentage of correctly classified peptides; *NP*, number of peptides.

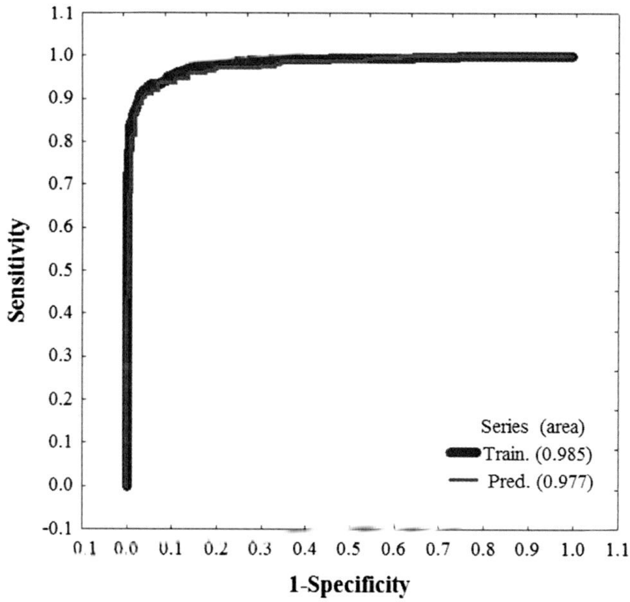

FIGURE 6.2 Receiver operating characteristic curves obtained for the multitasking chemoinformatic model.

6.3.2. Advantages and Limitations of the Present Model

One of the main advantages of our mtk-chemoinformatic model is its ability to simultaneously predict the anticancer activity against multiple cancer cell lines and the cytotoxicity against diverse healthy mammalian cells. This is possible because the molecular descriptors used to develop the model take into account both the peptide sequences and at least one of the elements of the experimental condition/ontology c_r: measures of the biological effects (m_e), and biological targets (b_t). Consequently, a defined peptide can appear more than once in the data set because of the different assay conditions under which the peptide was tested, and each time, the mtk-chemoinformatic model will yield a different classification/prediction result. Another important advantage is that all the molecular descriptors were generated from the peptide sequences, thus avoiding the high computational cost related to the optimization of three-dimensional structures. Therefore the model can be viewed as an alignment-free prediction tool.

Yet some limitations in the use of our model are to be referred. On the one hand, only linear peptides containing unmodified natural amino acids can be predicted. However, one should bear in mind that the chemical space embodying all the possible combinations of the 20 classical amino acids to form peptides is vast, and therefore, this aspect enhances even more the practical utility of our mtk-chemoinformatic model. On the other hand, here, only acetylation and amidation can be considered by the model as chemical modifications at the N-terminus and C-terminus, respectively. Nevertheless, it should be emphasized that acetylation and amidation are two of the most common chemical modifications present in the peptides.

6.3.3. Physicochemical Interpretations of the Molecular Descriptors

In drug discovery, there is a tendency for using computational models to perform virtual screening of large chemical libraries, including those derived from peptide research. In most cases, there is no attempt of gathering relevant physicochemical information from the models. This situation gets even more complicated for ANN or any other machine learning technique because there is no equation (in contrast with methods such as linear discriminant analysis or multiple linear regression) to point the analyst in the right direction regarding aspects of the molecular structure (encoded by the molecular descriptors) that should be varied to improve the biological effect (increase activity and/or diminish toxicity, among other properties). Therefore, computational models based on machine learning methods seem to be useless for rational drug design. Here, we propose a very simple approach with the aim of interpreting the different molecular descriptors of the mtk-chemoinformatic model from a physicochemical point of view. To accomplish this task, several steps were followed:

First, we examined the relative influences of the molecular descriptors in the model, which were obtained by the sensitivity analysis (Fig. 6.3). Then, we calculated the mean of each molecular descriptor in the model for both

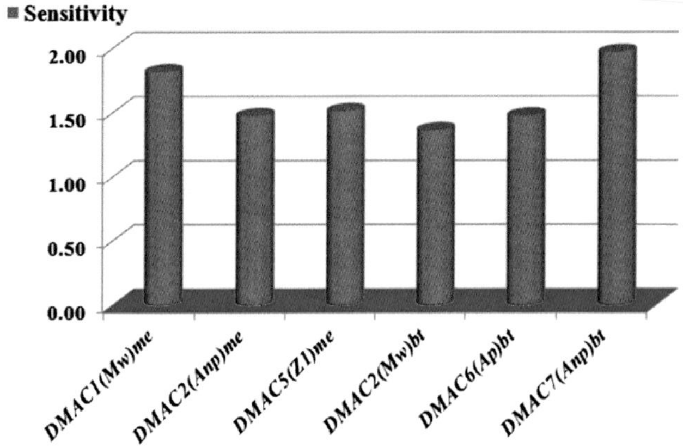

FIGURE 6.3 Sensitivity analysis performed with the aim of determining the relative importance of the molecular descriptors in the model.

TABLE 6.3 Means of the Different Molecular Descriptors That Entered in the Multitasking Chemoinformatic Model

	Means		
Descriptor	Positive (1)	Negative (−1)	Assumption
DMAC1(Mw)m_e	−232.26	−3394.01	Increase
DMAC2(Anp)m_e	−174.02	−1583.05	Increase
DMAC5(Z1)m_e	0.02	2.11	Decrease
DMAC2(Mw)b_t	−163.11	524.74	Decrease
DMAC6(Ap)b_t	−41.17	−205.84	Increase
DMAC7(Anp)b_t	22.41	−1782.34	Increase

positive and negative cases (Table 6.3). Finally, we compared the means within each molecular descriptor. Such comparisons allowed us to understand how the molecular descriptors should be varied to increase the anticancer activity and/or decrease the cytotoxicity of the peptides. It is important to emphasize here that such variations in the values of the molecular descriptors should be interpreted with caution because of the complex nonlinearity of our model. Thus, we will discuss only slight to middle variations in the values of the molecular descriptors.

Nonpolar area is perhaps the physicochemical property with the greatest importance. Indeed, by inspecting Fig. 6.3 and Table 6.3, one can see that the descriptor ***DMAC7(Anp)***$\boldsymbol{b_t}$ is the most influential variable in the mtk-chemoinformatic model, and it characterizes the increment of the nonpolar area of the amino acids that are placed at topological distance equal to 7, depending on the biological targets against which a peptide was tested. In convergence with ***DMAC7(Anp)***$\boldsymbol{b_t}$, the variable ***DMAC2(Anp)***$\boldsymbol{m_e}$ also expresses the increment of the nonpolar area, but only in regions where amino acids are separated at topological distance equal to 2, depending on the measure of biological effect.

Another property with remarkable impact on the biological effect of the peptides is the molar mass. In this sense, the descriptor ***DMAC1(Mw)***$\boldsymbol{m_e}$ (the second most important) characterizes the augmentation of the molar mass along the peptide sequence, and depending on the measure of biological effect. However, ***DMAC1(Mw)***$\boldsymbol{m_e}$ is constrained by ***DMAC2(Mw)***$\boldsymbol{b_t}$, which describes the diminution of the molar mass in regions where the amino acids are placed at topological distances equal to 2. It should be pointed out that ***DMAC2(Mw)*** $\boldsymbol{b_t}$ depends on the biological targets used in the assays, and it is the descriptor with the lowest influence in the model.

Finally, there are two more descriptors that deserve special attention. From one side, ***DMAC5(Z1)m$_e$*** depends on the measure of biological effect, indicating the decrease of the hydrophobicity-related properties in the peptide sequence where the amino acids are placed at topological distance equal to 5. This means that one of the amino acids in each pair should be hydrophobic, while the other amino acid should be hydrophilic. Therefore, ***DMAC5(Z1)m$_e$*** (the third most important descriptor) is a specific measure of the amphiphilicity of the peptides. On the other hand, as in the case of the nonpolar area (described by ***DMAC7(Anp)b$_t$*** and ***DMAC2(Anp)m$_e$***), the presence of polar areas seems to be a crucial aspect, which is characterized by the descriptor ***DMAC6(Ap)b$_t$***. In this context, ***DMAC6(Ap)b$_t$*** expresses the increment of the polar area in regions where the amino acids are placed at topological distance equal to 6, with dependence on the biological targets against which a defined peptide was assayed. This descriptor has the fourth greatest influence in the model.

The joint interpretation of all these molecular descriptors indicates that in order to increase the anticancer activity and diminish the cytotoxicity, peptides should be combinations of heavy amino acids exhibiting large nonpolar areas [tryptophan (W), phenylalanine (F), leucine (L), isoleucine (I), etc.], with medium-weight [valine (V)] and lightweight [alanine (A)] amino acids. Additionally, amino acids with a good balance of the different physicochemical properties must be present. This is the case of lysine (K), which also gives a cationic nature to the peptides, allowing them to disrupt the cellular membrane of the cancer cells, with the subsequent inhibition of the growth (Tyagi et al., 2015). The presence of hydrophilic amino acids such as arginine (R) and aspartic acid (D) is also necessary, which should appear dispersed along the peptide sequence.

6.3.4. Virtual Design and Screening of Peptide With Potential Anticancer Activities and Low Cytotoxicities

In peptide discovery, current computational models are used as "black boxes" with the sole purpose of performing predictions of peptides. Such virtual predictions are accurate to some extent, i.e., they have limited confidence. This fact can be explained because there are no universal descriptors, which means that the descriptors employed in any model consider only a reduced fraction of the vast chemical space characterized by a finite but huge diversity and complexity associated with the peptides. In this context, even those models that use combinations of different molecular descriptors constitute mere approximations of the real phenomena regarding the activities and toxicities of peptides. In this chapter, we present the first attempt devoted to designing peptides by using a computational model based on a machine learning method. In this subsection, we pretend to demonstrate that at least from a theoretical point of view, potent anticancer peptides also exhibiting low cytotoxicity against healthy cells can be designed.

TABLE 6.4 Peptides That Were Designed From the Physicochemical Interpretations of the Molecular Descriptors in the Multitasking Chemoinformatic Model

ID	Sequence	Molar Mass (Mw)
P1	VFKSHDRVKFKVHVKVHVKFKVRDHSKVF	3540
P2	VVKSHDRFKFKVHVKVHVKFKVRDHSKVF	3540
P3	VIKSHDRIKFKVHVKVHIKFKIRDHSKII	3528
P4	VFKSHDRIKFKVHVKVHIKFKIRDHSKII	3562
P5	VIKSHDRFKFKVHVKVHIKFKIRDHSKII	3562
P6	VIKSHDRIKFKVHIKVHVKFKIRDHSKII	3528
P7	VIKSHDRIKFKVHVKVHIKFKIRDHSKIF	3562
P8	VVKSHDRIKFKVHVKVHIKFKIRDHSKIF	3548
P9	VFKSHDRVKFKIHVKVHIKFKIRDHSKIF	3596
P10	VFKSHDRWKFKAHAKAHWKFKWRDHSKWW	3843
P11	VWKSHDRFKFKAHAKAHWKFKWRDHSKWW	3843
P12	VWKSHDRWKFKAHWKAHAKFKWRDHSKWW	3882

Bearing in mind all the ideas regarding the physicochemical interpretation (gathered from Fig. 6.3 and Table 6.3) of the different molecular descriptors, and with the purpose of illustrating the practical use of the mtk-chemoinformatic model as a tool for rational design and virtual screening, we created a chemical library formed by 12 peptides, each containing 29 amino acids (Table 6.4). Peptides were generated by considering those amino acids that exhibited the best balance among the different physicochemical properties characterized by the molecular descriptors. As mentioned in the first section of the present book chapter, the stability of the peptides is an aspect of crucial importance, and a peculiar detail of the designed peptides is that all of them have the amino acid valine (V) at the N-terminus, and valine is considered one of the amino acids strongly associated with increasing the half-life of peptides according to the N-end rule (Bachmair et al., 1986). In any case, in order to have a theoretical confirmation regarding the stability of the designed peptides, we performed predictions by using the ProtParam tool, which belong to the ExPASy web server (Gasteiger et al., 2003; Wilkins et al., 1999). ProtParam is able to estimate the half-life of peptides by considering three different biological systems: mammalian reticulocytes, yeast, and *Escherichia coli*. In addition, ProtParam determines the instability index. In all cases, the peptides were classified to exhibit high stability with half-life values

>100 h, >20 h, and >10 h for the aforementioned biological systems, respectively.

The mtk-chemoinformatic model was used to predict the anticancer activity and cytotoxicity of the designed peptides. A summary of the results of the predictions are represented in Table 6.5, while the file containing the data of the designed peptides can be found in the supplementary material (**SM5.xlsx**) via the companion website associated with this book. From Table 6.5, it can be seen that the 12 designed peptides were predicted to have potent anticancer activity against all the cancer cell lines ($IC_{50} \leq 25$ μM).

TABLE 6.5 Summary of the Predictions Performed by the Multitasking Chemoinformatic Model

Measure of Biological Effect (m_e)	Biological Target (b_t)	Pred-$BE_q(c_r)$[a,b]
IC_{50} (μM)	Lung cancer cells (NCI-H128)	1
	Prostate cancer cells (PC-3)	1
	Lung cancer cells (NCI-H69)	1
	Lung cancer cells (NCI-H146)	1
	Cervical cancer cells (HeLa)	1
	Colorectal cancer cells (SW-480)	1
	Lung cancer cells (H-1299)	1
	Breast cancer cells (MDA-MB-361)	1
	Leukemia cancer cells (K-562)	1
	Breast cancer cells (MCF-7)	1
	Colorectal cancer cells (SW-1116)	1
	Skin cancer cells (BMKC)	1
	Renal cancer cells (A498)	1
	Leukemia cancer cells (Jurkat)	1
	Skin cancer cells (A375)	1
	Skin cancer cells (B-16)	1
	Lung cancer cells (A549)	1
	Rhabdomyosarcoma cancer cells (RD)	1
	Colorectal cancer cells (HCT-116)	1
	Leukemia cancer cells (CCRF-CEM)	1

TABLE 6.5 Summary of the Predictions Performed by the Multitasking Chemoinformatic Model—cont'd

Measure of Biological Effect (m_e)	Biological Target (b_t)	Pred-$BE_q(c_r)$[a,b]
CC_{50} (μM)	MARC-145 cells	1
	Murine macrophage cells (Raw 264.7)	1
	Mouse fibroblast cells (NIH 3T3)	1
	Human monocytic THP-1 cells	1
	Murine fibroblast cells (L929)	1
	Human umbilical vein endothelial cells	1
	Rat epithelial cells (IEC-6, CRL-1592)	1
	CEM-SS cells	−1
	Murine splenocytes	−1
	Madin–Darby canine kidney cells	−1
HC_{50} (μM)	Sheep erythrocytes	1
	Rabbit erythrocytes	1
	Human erythrocytes	1
	Rat erythrocytes	1
	Mouse erythrocytes	1

[a]*Predicted value of the categorical variable of biological effect [$BE_q(c_r)$] of the peptide.*
[b]*The multitasking chemoinformatic model yielded the same predictions for all the designed peptides.*

In terms of cytotoxicity, each of these peptides was classified as positive (noncytotoxic, $CC_{50} \geq 80.37$ μM) for 7 of 10 healthy mammalian cells other than erythrocytes. Finally, for the case of the five different types of erythrocytes, all the peptides were classified as positive (nontoxic/nonhemolytic, $HC_{50} \geq 105.7$ μM). The results of the predictions suggest that the peptides may be used in future biological tests focused on the assessment of their anticancer activities and cytotoxicities.

6.4. CONCLUSIONS

Anticancer peptides constitute hopeful alternatives in chemotherapy to forward the improvement of the quality of life of human beings. Due to the explosion of chemical and biological data associated with peptide discovery, computational models should be used beyond the classical task of performing virtual screening of peptides with unspecified anticancer activity, and more attention should be paid to the use of such models as knowledge generators, i.e., tools able to design peptides with the desired biological profiles. The present mtk-chemoinformatic model represents an attempt to achieve such a goal. Our model displayed a high performance for classifying/predicting many peptides assayed against multiple cancer cell lines and/or diverse mammalian cells. From the interpretation of the molecular descriptors in the model, it was possible to design 12 new peptides not reported in our database. The theoretical predictions performed by the mtk-chemoinformatic model yielded encouraging results regarding the fact that the designed peptides may have high and versatile anticancer activity, and low cytotoxicity. The computational methodology explained here can be generalized to other fields of research in drug discovery. Our mtk-chemoinformatic model opens new horizons for the rational design and efficient virtual screening of wide-spectrum anticancer peptides with low cytotoxic effects.

ACKNOWLEDGMENTS

The authors are grateful for the joint financial support given by the Portuguese Fundação para a Ciência e a Tecnologia (FCT/MEC) and FEDER (Projects No. UID/QUI/50006/2013 and POCI/01/0145/FEDER/007265).

REFERENCES

Agrawal, P., Bhalla, S., Usmani, S.S., Singh, S., Chaudhary, K., Raghava, G.P., Gautam, A., 2016. CPPsite 2.0: a repository of experimentally validated cell-penetrating peptides. Nucleic Acids Res. 44, D1098–D1103.

Bachmair, A., Finley, D., Varshavsky, A., 1986. In vivo half-life of a protein is a function of its amino-terminal residue. Science 234, 179–186.

Chaudhary, K., Kumar, R., Singh, S., Tuknait, A., Gautam, A., Mathur, D., Anand, P., Varshney, G.C., Raghava, G.P., 2016. A web server and mobile app for computing hemolytic potency of peptides. Sci. Rep. 6, 22843.

Chen, F., Ma, B., Yang, Z.C., Lin, G., Yang, D., 2012. Extraordinary metabolic stability of peptides containing alpha-aminoxy acids. Amino Acids 43, 499–503.

Chen, W., Ding, H., Feng, P., Lin, H., Chou, K.C., 2016. iACP: a sequence-based tool for identifying anticancer peptides. Oncotarget 7, 16895–16909.

Cheng, H., Force, T., 2010. Molecular mechanisms of cardiovascular toxicity of targeted cancer therapeutics. Circ. Res. 106, 21–34.

Chu, H.L., Yip, B.S., Chen, K.H., Yu, H.Y., Chih, Y.H., Cheng, H.T., Chou, Y.T., Cheng, J.W., 2015. Novel antimicrobial peptides with high anticancer activity and selectivity. PLoS One 10, e0126390.

de la Fuente-Nunez, C., Reffuveille, F., Haney, E.F., Straus, S.K., Hancock, R.E., 2014. Broad-spectrum anti-biofilm peptide that targets a cellular stress response. PLoS Pathog. 10, e1004152.

Dennison, S.R., Harris, F., Bhatt, T., Singh, J., Phoenix, D.A., 2010. A theoretical analysis of secondary structural characteristics of anticancer peptides. Mol. Cell. Biochem. 333, 129–135.

E-Kobon, T., Thongararm, P., Roytrakul, S., Meesuk, L., Chumnanpuen, P., 2016. Prediction of anticancer peptides against MCF-7 breast cancer cells from the peptidomes of *Achatina fulica* mucus fractions. Comput. Struct. Biotechnol. J. 14, 49–57.

Gaschen, B., Kuiken, C., Korber, B., Foley, B., 2001. Retrieval and on-the-fly alignment of sequence fragments from the HIV database. Bioinformatics 17, 415–418.

Gasteiger, E., Gattiker, A., Hoogland, C., Ivanyi, I., Appel, R.D., Bairoch, A., 2003. ExPASy: the proteomics server for in-depth protein knowledge and analysis. Nucleic Acids Res. 31, 3784–3788.

Gautam, A., Chaudhary, K., Singh, S., Joshi, A., Anand, P., Tuknait, A., Mathur, D., Varshney, G.C., Raghava, G.P., 2014. Hemolytik: a database of experimentally determined hemolytic and non-hemolytic peptides. Nucleic Acids Res. 42, D444–D449.

Gentilucci, L., De Marco, R., Cerisoli, L., 2010. Chemical modifications designed to improve peptide stability: incorporation of non-natural amino acids, pseudo-peptide bonds, and cyclization. Curr. Pharm. Des. 16, 3185–3203.

Gogoladze, G., Grigolava, M., Vishnepolsky, B., Chubinidze, M., Duroux, P., Lefranc, M.P., Pirtskhalava, M., 2014. DBAASP: database of antimicrobial activity and structure of peptides. FEMS Microbiol. Lett. 357, 63–68.

Gorris, H.H., Bade, S., Rockendorf, N., Albers, E., Schmidt, M.A., Franek, M., Frey, A., 2009. Rapid profiling of peptide stability in proteolytic environments. Anal. Chem. 81, 1580–1586.

Gupta, S., Kapoor, P., Chaudhary, K., Gautam, A., Kumar, R., Raghava, G.P., 2013. In silico approach for predicting toxicity of peptides and proteins. PLoS One 8, e73957.

Hajisharifi, Z., Piryaiee, M., Mohammad Beigi, M., Behbahani, M., Mohabatkar, H., 2014. Predicting anticancer peptides with Chou's pseudo amino acid composition and investigating their mutagenicity via Ames test. J. Theor. Biol. 341, 34–40.

Hein, K.Z., Takahashi, H., Tsumori, T., Yasui, Y., Nanjoh, Y., Toga, T., Wu, Z., Grotzinger, J., Jung, S., Wehkamp, J., Schroeder, B.O., Schroeder, J.M., Morita, E., 2015. Disulphide-reduced psoriasin is a human apoptosis-inducing broad-spectrum fungicide. Proc. Natl. Acad. Sci. U. S. A. 112, 13039–13044.

Hill, T., Lewicki, P., 2006. Statistics Methods and Applications. A Comprehensive Reference for Science, Industry and Data Mining. StatSoft, Tulsa.

Illouz, F., Braun, D., Briet, C., Schweizer, U., Rodien, P., 2014. Endocrine side-effects of anti-cancer drugs: thyroid effects of tyrosine kinase inhibitors. Eur. J. Endocrinol. 171, R91–R99.

Kapoor, P., Singh, H., Gautam, A., Chaudhary, K., Kumar, R., Raghava, G.P., 2012. TumorHoPe: a database of tumor homing peptides. PLoS One 7, e35187.

Karpel-Massler, G., Horst, B.A., Shu, C., Chau, L., Tsujiuchi, T., Bruce, J.N., Canoll, P., Greene, L.A., Angelastro, J.M., Siegelin, M.D., 2016. A synthetic cell-penetrating dominant-negative ATF5 peptide exerts anti-cancer activity against a broad spectrum of treatment resistant cancers. Clin. Cancer Res. 22, 4698–4711.

Kumar, R., Chaudhary, K., Sharma, M., Nagpal, G., Chauhan, J.S., Singh, S., Gautam, A., Raghava, G.P., 2015. AHTPDB: a comprehensive platform for analysis and presentation of antihypertensive peptides. Nucleic Acids Res. 43, D956–D962.

Maher, S., McClean, S., 2006. Investigation of the cytotoxicity of eukaryotic and prokaryotic antimicrobial peptides in intestinal epithelial cells in vitro. Biochem. Pharmacol. 71, 1289–1298.

Maher, S., McClean, S., 2008. Melittin exhibits necrotic cytotoxicity in gastrointestinal cells which is attenuated by cholesterol. Biochem. Pharmacol. 75, 1104–1114.

Martinez-Lacaci, I., Garcia Morales, P., Soto, J.L., Saceda, M., 2007. Tumour cells resistance in cancer therapy. Clin. Transl. Oncol. 9, 13–20.

McGuire, S., 2016. World Cancer Report 2014. Geneva, Switzerland: World Health Organization, International Agency for Research on Cancer, WHO Press, 2015. Adv. Nutr. 7, 418–419.

Mehta, D., Anand, P., Kumar, V., Joshi, A., Mathur, D., Singh, S., Tuknait, A., Chaudhary, K., Gautam, S.K., Gautam, A., Varshney, G.C., Raghava, G.P., 2014. ParaPep: a web resource for experimentally validated antiparasitic peptide sequences and their structures. Database (Oxford).

Nguyen, L.T., Haney, E.F., Vogel, H.J., 2011. The expanding scope of antimicrobial peptide structures and their modes of action. Trends Biotechnol. 29, 464–472.

Paul, S.M., Mytelka, D.S., Dunwiddie, C.T., Persinger, C.C., Munos, B.H., Lindborg, S.R., Schacht, A.L., 2010. How to improve R&D productivity: the pharmaceutical industry's grand challenge. Nat. Rev. Drug Discov. 9, 203–214.

Pearson, K., 1895. Notes on regression and inheritance in the case of two parents. Proc. R. Soc. Lond. 58, 240–242.

Pirtskhalava, M., Gabrielian, A., Cruz, P., Griggs, H.L., Squires, R.B., Hurt, D.E., Grigolava, M., Chubinidze, M., Gogoladze, G., Vishnepolsky, B., Alekseev, V., Rosenthal, A., Tartakovsky, M., 2016. DBAASP v.2: an enhanced database of structure and antimicrobial/cytotoxic activity of natural and synthetic peptides. Nucleic Acids Res. 44, D1104–D1112.

Romero-Duran, F.J., Alonso, N., Yanez, M., Caamano, O., Garcia-Mera, X., Gonzalez-Diaz, H., 2016. Brain-inspired cheminformatics of drug-target brain interactome, synthesis, and assay of TVP1022 derivatives. Neuropharmacology 103, 270–278.

Ruiz-Blanco, Y.B., Paz, W., Green, J., Marrero-Ponce, Y., 2015. ProtDCal: a program to compute general-purpose-numerical descriptors for sequences and 3D-structures of proteins. BMC Bioinform. 16, 162.

Speck-Planche, A., Cordeiro, M.N.D.S., 2014. Chemoinformatics for medicinal chemistry: in silico model to enable the discovery of potent and safer anti-cocci agents. Future Med. Chem. 6, 2013–2028.

Speck-Planche, A., Kleandrova, V.V., Ruso, J.M., Cordeiro, M.N.D.S., 2016. First multitarget chemo-bioinformatic model to enable the discovery of antibacterial peptides against multiple gram-positive pathogens. J. Chem. Inf. Model. 56, 588–598.

Statsoft-Team, 2001. STATISTICA. v6.0. Data Analysis Software System, Tulsa.

Tenorio-Borroto, E., Penuelas-Rivas, C.G., Vasquez-Chagoyan, J.C., Castanedo, N., Prado-Prado, F.J., Garcia-Mera, X., Gonzalez-Diaz, H., 2014. Model for high-throughput screening of drug immunotoxicity – study of the anti-microbial G1 over peritoneal macrophages using flow cytometry. Eur. J. Med. Chem. 72, 206–220.

Todeschini, R., Consonni, V., 2000. Handbook of Molecular Descriptors. WILEY-VCH Verlag GmbH, Weinheim, New York, Chichester, Brisbane, Singapore, Toronto.

Torino, F., Barnabei, A., Paragliola, R.M., Marchetti, P., Salvatori, R., Corsello, S.M., 2013. Endocrine side-effects of anti-cancer drugs: mAbs and pituitary dysfunction: clinical evidence and pathogenic hypotheses. Eur. J. Endocrinol. 169, R153–R164.

Torre, L.A., Bray, F., Siegel, R.L., Ferlay, J., Lortet-Tieulent, J., Jemal, A., 2015. Global cancer statistics, 2012. CA Cancer J. Clin. 65, 87–108.

Tyagi, A., Kapoor, P., Kumar, R., Chaudhary, K., Gautam, A., Raghava, G.P., 2013. In silico models for designing and discovering novel anticancer peptides. Sci. Rep. 3, 2984.

Tyagi, A., Tuknait, A., Anand, P., Gupta, S., Sharma, M., Mathur, D., Joshi, A., Singh, S., Gautam, A., Raghava, G.P., 2015. CancerPPD: a database of anticancer peptides and proteins. Nucleic Acids Res. 43, D837–D843.

Verges, B., Walter, T., Cariou, B., 2014. Endocrine side effects of anti-cancer drugs: effects of anticancer targeted therapies on lipid and glucose metabolism. Eur. J. Endocrinol. 170, R43–R55.

Wilkins, M.R., Gasteiger, E., Bairoch, A., Sanchez, J.C., Williams, K.L., Appel, R.D., Hochstrasser, D.F., 1999. Protein identification and analysis tools in the ExPASy server. Methods Mol. Biol. 112, 531–552.

Zhao, H., Zhou, J., Zhang, K., Chu, H., Liu, D., Poon, V.K., Chan, C.C., Leung, H.C., Fai, N., Lin, Y.P., Zhang, A.J., Jin, D.Y., Yuen, K.Y., Zheng, B.J., 2016. A novel peptide with potent and broad-spectrum antiviral activities against multiple respiratory viruses. Sci. Rep. 6, 22008.

Chapter 7

Flavonoids From Asteraceae as Multitarget Source of Compounds Against Protozoal Diseases

É.B.V.S. Cavalcanti[1], V. de Paulo Emerenciano[2], L. Scotti[1], M.T. Scotti[1]
[1]*Federal University of Paraíba, João Pessoa, Paraíba, Brazil;* [2]*University of São Paulo, São Paulo, SP, Brazil*

7.1. INTRODUCTION

Brazil has great social problems due to parasitic infections such as leishmaniasis, trypanosomiasis, schistosomiasis, and malaria, which are among the most neglected despite the social issues arising and also resulting from these diseases. Furthermore, the emergence of drug-resistant strains, in addition to the side effects associated with the use of the current antiparasitic drugs, indicate the urgent need for new and alternative chemotherapies (Lindoso and Lindosa, 2009).

The search for new drugs against these diseases is urgently needed, and natural sources such as plants with its various secondary metabolites may play an important role (Schmidt et al., 2012). Several scientific studies have demonstrated the pharmacological efficiency of flavonoids and other natural products in the treatment of various diseases. In this context, chemical families that can be found in the literature are sesquiterpene lactones with promising antiprotozoal activity, in particular trypanocidal activity (Karioti et al., 2009).

Plants are a valuable source of inspiration for the development of therapeutic agents (Mishra and Tiwari, 2011). It is estimated that the current market of plant-derived drugs handles more than $20 billion and continues to grow. However, only 10–15% of plant species have been exploited for the development of clinically important drugs. Therefore, natural products have been shown to be a promising alternative in treating various diseases (Filho and Yunes, 1998; Montanari and Bolzani, 2001). Studies of rational planning based

Multi-Scale Approaches in Drug Discovery. http://dx.doi.org/10.1016/B978-0-08-101129-4.00007-2

on these compounds generate results that can feature a pharmacophore providing a prototype of this drug.

The search for new ligands derived from natural products as a base for new antiprotozoal agents is a viable strategy for these products that account for more than 50% of novel ligands, either directly or indirectly (synthetic origin compounds inspired by the natural product structure) of the medicines approved in the period 1981–2014 (Newman and Cragg, 2016).

The data organization, i.e., the structure and the species from which it was extracted, assists in obtaining information more rapidly and with lower cost at all levels in terms of time, materials, equipment and labor. From these databases, one can perform queries, reports, or even use the structures for various theoretical studies, and combine the data with other databases available in the literature (Gasteiger, 2003; Gasteiger and Engel, 2003).

Virtual screening is an increasingly utilized approach with the increasing number of publications and citations (Fig. 7.1). It is also being used in the pharmaceutical industry, as a consequence of the increased number of databases containing high volumes of structural and biological information of many chemicals/drugs. The virtual screening studies allow to rank or filter a number of databases of compounds using one or more computational procedures, which may be used to prioritize the compounds that should be synthesized or purchased (Gasteiger, 2003; Gasteiger and Engel, 2003).

The use of complex networks as statistical tools to study the link between drugs and targets is an innovative approach in the field of public health (Sousa and Ruiz, 2015). The results of the botanical occurrence of secondary metabolites organized in a database, and the relationships between chemical structure and antiprotozoan activity of these compounds are of great value to minimize time and financial costs for bioprospecting, particularly, obtaining new drugs and valuating the biodiversity as a source of structural proposals for

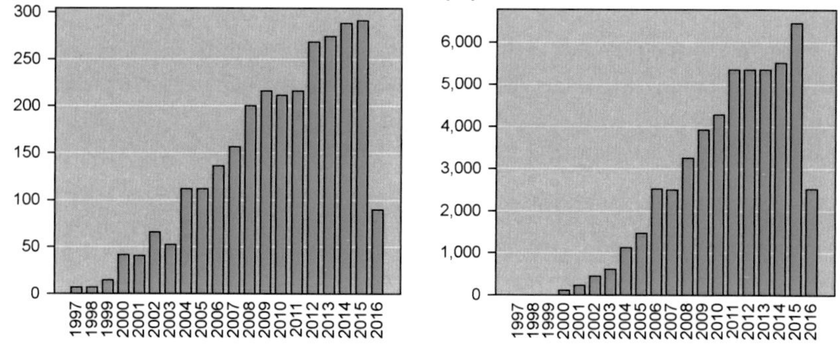

FIGURE 7.1 Number of publications (A) and citations (B) to the term "virtual screening" in titles. *Web of science.*

potential new ligands against neglected diseases such as leishmaniasis, trypanosomiasis, schistosomiasis, and malaria.

7.2. NEGLECTED DISEASES CAUSED BY PROTOZOA

There are times when humanity is beset by poverty-related diseases, which proliferate, especially in environments marked by social exclusion (Andrade and Rocha, 2015). More than 1 billion people live with less than US $2 a day, either in Brazil or other countries in Latin America and the Caribbean, Africa, Asia, and also in the United States and some European countries. They are mainly in the countryside, in urban areas of extreme poverty, and in conflict regions. They suffer all kinds of lack—of drinking water, education, sanitation, housing, and access to health treatments—and are the main victims of the neglected diseases (Assad, 2010).

Neglected diseases are medical conditions that prevail not only in situations of poverty but also contribute to the maintenance of unequal framework, as they represent a strong barrier to the development of countries (Ministério da Saúde, 2010b). These are treatable and curable diseases that mainly affect people with few financial resources, and rightfully so, do not arouse the interest of the pharmaceutical industry (Equipe Médica do MSF, 2012). The methods of treatment and diagnosis of these diseases are old and inadequate and require investment in research and development to become simpler and more effective (Equipe Médica do MSF, 2012).

Examples of neglected diseases include dengue, American trypanosomiasis (Chagas disease), schistosomiasis, leprosy, leishmaniasis, malaria, and tuberculosis (Ministério da Saúde, 2010b). According to the World Health Organization (WHO), these include other diseases such as trachoma, Buruli ulcer, African trypanosomiasis (sleeping sickness), dracunculiasis, cysticercosis, lymphatic filariasis, onchocerciasis; yaws, geohelminths, hydrophobia (rabies), echinococcosis, and fascioliasis (Andrade and Rocha, 2015).

With 1.5 billion people affected by neglected diseases in a universe of 149 countries, the WHO requested the countries to invest more in combating a number of neglected tropical diseases (Agência Brasil, 2015) that kill about 534,000 people in the world each year (Center for Disease Control and Prevention, 2016).

7.2.1. Leishmaniasis

Leishmaniasis is an infectious disease prevalent in Europe, Africa, Asia, and the Americas. It is endemic in 88 countries and kills thousands and debilitates millions of people every year (Mello et al., 2014).

Leishmaniasis is caused by more than 20 species of *Leishmania* protozoan and are transmitted to humans by about 30 species of phlebotomus vectors. They are classified into three main manifestations. (1) Visceral leishmaniasis or kala-azar is the most severe form; is systemic; can reach the liver, spleen, and bone marrow; is caused by *Leishamania donovani*, *Leishmania infantum*, and *Leishmania chagasi*; and has a high mortality rate if untreated. (2) Cutaneous leishmaniasis is characterized by chronic skin ulcers, develops at the site of the bite of the insect vector, and is caused by *Leishmania mexicana*, *Leishmania major*, and *Leishmania tropica*. Mucocutaneous leishmaniasis is generally caused by *Leishmania braziliensis* and destroys the nasal and oral mucosa (Soares-Bezerra et al., 2004).

Leishmania is a genus of protozoan trypanosome belonging to the family of obligate intracellular parasites of the mononuclear phagocyte system cells, with two main forms: the flagellated promastigote, found in the insect vector of the digestive tract, and the non-flagellated form or amastigotes, observed in the tissues of vertebrate hosts (Ministério da Saúde, 2010b).

The vectors of leishmaniasis are insects called phlebotomine sand flies, belonging to the order Diptera, family Psychodidae, subfamily Phlebotominae, and gender Lutzomyia, popularly known, depending on geographic location, as mosquito palha, tatuquira, birigui, among others. In Brazil, the main species involved in the transmission of the disease are *Lutzomyia flaviscutellata*, *Lutzomyia whitmani*, *Lutzomyia umbratilis*, *Lutzomyia intermedia*, *Lutzomyia wellcomei*, and *Lutzomyia migonei*. Phlebotomine sand flies measure 2–3 mm long and because of their small size are able to pass through the meshes of the nets and screens. They are yellowish or grayish and their wings are open when they are at rest (Ministério da Saúde, 2010a,b). The infected insect releases the promastigote form of the parasite that is inoculated into the vertebrate, engulfed by macrophages, and then, in the raw state amastigoste form (Ministério da Saúde, 2016).

The sources of infection of leishmaniasis are mainly wild animals and insect phlebotomine sand flies; however, the host can also be the domestic dog. In cutaneous leishmaniasis, animals that act as reservoirs are wild rodents, anteaters, and sloths. In visceral leishmaniasis, the main source of infection is the hoary fox (Ministério da Saúde, 2010b).

It is estimated that more than 12,000,000 million people are infected worldwide, and Brazil, alongside Bangladesh, India, Ethiopia, Kenya, and Sudan are the most affected, with 2,000,000 new cases, 1,500,000 cases of cutaneous leishmaniasis and 500,000 of visceral leishmaniasis. In Brazil, leishmaniasis is present in 19 states, and more than 90% of human cases of the disease are concentrated in the Northeast; however, there are still important centers in the Midwest, North, and South (Gil et al., 2008).

The treatment of leishmaniasis has been ineffective; it is administered in injectable form, which is painful and leads to many patients abandoning treatment. The drugs used in the treatment have one or more limitations such as high cost, toxicity issues, or resistance of parasites to drugs (Mishra et al., 2009).

The first-line drugs for the treatment of leishmaniasis are: complex ion pentavalent antimony, meglumine antimoniate (Glucantime) and sodium stibogluconate (Pentostam). These compounds were introduced in 1945 and are still effective for some forms of leishmaniasis. However, high doses are required, the drugs have high toxicity, and the patient must be hospitalized for 3–4 weeks for parenteral administration (Bastos et al., 2012).

The therapeutic dose and time vary with the forms of the disease and the severity of symptoms. Moreover, these drugs have side effects. Glucantime induces cardiac arrhythmias and is contraindicated in patients taking β-blockers and antiarrhythmic drugs with renal or hepatic failure and in pregnant women in the first two trimesters of pregnancy, which advocates the importance of the correct assessment of the conditions, generally before starting the treatment (Martins and Lima, 2013). The intralesional injection of sodium stibogluconate is very painful and can lead to the formation of hyperpigmented scars (Bastos et al., 2012).

According to Medda et al. (1999), the second possible treatment is the use of pentamidine, paromomycin, miltefosine, and amphotericin B administered parenterally. Amphotericin B formulations show greater efficacy in treating visceral leishmaniasis when associated with lipids; however, the cost becomes unfavorably high when compared with other treatments.

Another important strategy is the use of cryotherapy, which has the advantages of good tolerability and low cost. Studies have shown that its effectiveness is greatly enhanced when combined with antimony therapy (Asilian et al., 2004; Salmanpour et al., 2006).

Recent studies have also investigated the efficacy of photodynamic therapy as an alternative therapy that involves intravenous administration of a photosensitizer, which binds to low-density lipoproteins in the bloodstream (Bastos et al., 2012). Since malignant cells have a higher quantity of low-density lipoproteins, the photosensitizer is concentrated in these tissues and then activated by light at a certain wavelength, selectively, destroying the infected cells (Bastos et al., 2012). Some photosensitizers are derived from porphyrins, and their antiparasite effects are probably due to the killing of infected host cells, not by killing the parasite.

The investigation of a small series of cationic photosensitizers suggests that it may be possible to photodynamically inactivate macrophages infected with *Leishmania* promastigotes. Thus, further studies are being developed to identify the structure of these substances by high activity and selectivity against *Leishmania* (Bastos et al., 2012).

One of the features in the process of drug development is target identification in a biological pathway. In theory, during the identification of a target in a pathogen, it is important that the putative target be either absent in the host or substantially different from the homologous host, so that it can be exploited as a drug target. Second, the target selected should be absolutely necessary for the survival of the pathogen. It is also important to consider the stage of the life cycle of the pathogen in which the target gene is expressed. It is crucial to look at the biochemical properties of the protein; it should have a small molecule—binding pocket, so that specific inhibitors can be designed, and if the target protein is an enzyme, its inhibition should lead to loss in cell viability (Chawla and Madhubala, 2010).

Some metabolic pathways are essential for the life of the parasite, and these reactions usually involve extremely complex chemical mechanisms. A very large number of reactions are efficiently catalyzed by specific enzymes. Research toward the discovery of new anti-*Leishmania* drugs involves the search for new biochemical targets related to the mechanism of defense; metabolism of RNA, DNA, glucose, sterols, fatty acids, purine pathway, and nucleotides; or any biochemical route of the parasite that can be attacked by the drug and is safe for the host (Scotti et al., 2015).

Natural products have been a rich source of compounds with anti-leishmanial activity (Williams et al., 2003). Some of the main enzymatic targets studied in the development of new drugs for the treatment of leishmaniasis, using natural products as inhibitors are as follows.

7.2.1.1. Adenine Phosphoribosyltransferase (ID PDB 1QB7)

This enzyme is found exclusively in *Leishmania* parasites (6-aminopurine metabolism), playing a paramount role in purine metabolism in this genus by transforming 6-aminopurines into 6-oxypurines (Barros-Alvarez et al., 2014; Boitz et al., 2012; Boitz and Ullman, 2013).

7.2.1.2. N*-myristoyltransferase (ID PDB 2WUU)*

N-myristoyltransferase (NMT) is a ubiquitous eukaryotic enzyme responsible for co- and posttranslational attachment of myristate (14:0) to an N-terminal glycine of substrate proteins, via an amide bond, and essential for viability in protozoan parasites with functions in multiple cellular processes. Protein *N*-myristoylation is important for targeting proteins to membrane sites; the enzyme catalyzes the attachment of the myristate to the amino-terminal glycine residue (Brannigan et al., 2010; Goldston et al., 2014). NMTs have been well characterized in a range of eukaryotes, being a suitable, potential, and a valid target for antifungal, anti-*Leishmania*, and antimalaria drug development (Tate et al., 2014; Rackham et al., 2014; Wright et al., 2014; Goncalves et al., 2012).

7.2.1.3. Dihydroorotate Dehydrogenase (ID PDB 3C61)

Pyrimidine nucleotides are essential for all forms of life; they serve as RNA and DNA precursors, in signaling, in cell membrane assembly, and in phospholipid, complex lipid, and glycoconjugate biosynthesis. Their biosynthetic pathway involves six enzymatic steps leading to the synthesis of uridine 5-monophosphate (French et al., 2011). In the same study, various polyphenol ligands, the chalcone crotaorixin, the flavonoid sophoraflavanone

G, the lignin aristolignin, the coumarins mammea A/AA and mammea B/BA, and the cannabinoid 4-acetoxy-2-geranyl-5-hydroxy-3-*n*pentylphenol showed promising docking with *L. major* dihydroorotate dehydrogenase (Ogungbe et al., 2014).

7.2.1.4. Peptidylprolyl Cis-trans Isomerase (Cyclophilin) (ID PDB 3EOV)

Three classes of structurally unrelated peptidylprolyl cis/trans isomerases catalyze the otherwise slow interconversion of *cis* and *trans* conformations of the peptidyl–prolyl bond: FK506-binding proteins, cyclophilins, and parvulins and catalyze the *cis-trans* isomerization of proline imidic peptide bonds in oligopeptides (Moore and Potter, 2013; Venugopal et al., 2009). Kitamura et al. (2013) studied antioxidant activities of various flavonoides (flavanones, chalcone, and isoflavone) by using recombinant *Escherichia coli* cells.

7.2.1.5. Arginase (ID PDB 4ITY)

Polyamines (PAs) are essential for cell proliferation and trypanothione production, which is involved in the detoxification of reactive oxygen species (Colotti and Ilari, 2011). In *Leishmania*, arginase produces L-ornithine, which is then decarboxylated by ornithine decarboxylase (ODC) to generate putrescine, which continues down the PA synthesis pathway (Da Silva et al., 2012). Studies of mutant *L. mexicana* and *L. major* parasites (lacking arginase) confirmed that the pathway is essential for parasite viability and infectivity (Da Silva et al., 2012; Iniesta et al., 2001).

7.2.1.6. Phosphodiesterase B1

It catalyzes cAMP hydrolysis to AMP, allowing AMP recycling for use in its several biochemical pathways. A study by Ogungbe et al. (2014) also found *L. major* phosphodiesterase B (LmajPDEB1) inhibitors. The best compounds were four stilbenoids (which also demonstrated anti-*Plasmodium* activity), and the antiplasmodial flavonoid artonin F (Namdaung et al., 2006).

7.2.1.7. Pteridine Reductase 1

Enzymes involved in the provision and use of reduced folate such as dihydrofolate reductase (DHFR), and thymidylate synthase (TS) are valued drug targets for the treating cancer, bacterial infections, and parasites such as malaria (Ogungbe et al., 2014). DHFR catalyzes a two-step reduction of folate to tetrahydrofolate, which is transformed to *N*5,*N*10-methylene tetrahydrofolate, and is used by TS in the conversion of 2′-deoxyuridine-5′-monophosphate to 2′-deoxythymidine-5′-monophosphate. Inhibition of DHFR or TS impairs DNA replication, and generally results in cell death. In the docking study conducted by Ogungbe et al. (2014), numerous coumarins, flavonoids, isoflavonoids, and lignans docked selectively to *L. major* DHFR1 (Cavazzuti et al., 2008).

7.2.2. American Trypanosomiasis

Known as Chagas disease, American trypanosomiasis is a parasitic infection caused by *Trypanosoma cruzi*, a protozoan whose life cycle includes the obligatory passage through various mammalian hosts, which are infected by an insect vector: the barber. This disease can also be considered a result of anthropozoonosis changes produced by human beings in environmental and economic inequalities (Fundação Osvaldo Cruz, 2008).

In Brazil, Chagas disease is a highly prevalent endemic disease in Mexico, Central America and South America. It is estimated that the number of infected people is 16–18 million, with over 100 million considered at risk of infection (Tasdemir et al., 2006). Although Brazil has drastically reduced the numbers of infections due to Chagas disease in recent decades, between 150 and 200 new cases are still reported annually. According to the Surveillance Secretariat of Health of the Ministry of Health, the disease is controlled in the South, Southeast, and Midwest, and there are rare cases in the Northeast (Fiocruz, 2012).

T. cruzi is a flagellated protozoan of the order Kinetoplastida, family trypanosomes, characterized by the existence of a single flagellum and kinetoplast, an organelle containing DNA and located in mitochondria. This protozoa can be transmitted to humans in the following ways: vector (classical), transfusion (reduced due to the sanitary control of hemoderivatives and hemocomponents), congenital (transplacental), accidental (accidents in laboratories), oral (contaminated food), and transplants. In the barber vector transmission, the insect bites, ingests blood, its infected feces are deposited near the bite wound. Scratching the site causes the T. cruzi trypomastigotes in the feces to enter the host through the wound. Historically, this type of transmission has been the main cause of Chagas disease in Brazil (Anvisa, 2008).

There are two phases of the disease: the acute stage that appears soon after infection and the chronic stage that appears after a silent and asymptomatic period that can last 10–20 years (Correa et al., 2010). The incubation period of Chagas disease varies according to the path of transmission, with 5–15 days in the vector, 30–40 days in blood transfusion, from the fourth to the ninth month of pregnancy in the placenta, and about 7–22 days oral route. The clinical condition is characterized by prolonged fever, headache, swelling of the face or limbs, spots on the skin, enlarged liver or spleen, and acute heart disease, among others (Anvisa, 2008).

The chronic phase is characterized by progressive inflammation of the heart muscle (myocarditis chagasica), which produces destruction of cardiac fibers and fibrosis in multiple areas of the myocardium (Correa et al., 2010). Confirmation of the disease is made by parasitological examination and serological tests as per medical advice (Anvisa, 2008).

For the treatment of Chagas disease two drugs are used in clinical practice: nifurtimox nitrofuran (Lampit/Bayer 2502), 1,1-dioxide tetrahydro-3-methyl-4 [(5-nitrofurfurilideno) amino]-2H-1,4-triazine, and benznidazole (N-benzyl-2-nitro-1-imidazolacetamide, a nitroimidazole derivative). The first is no longer used in Brazil due to its significant toxic effects, and benznidazole has been used to date, but several authors also point out significant toxicity (Castro, 2000; Castro et al., 2006).

Because of the low tolerability and limited accessibility of current chemotherapeutic regimens for all the neglected parasitic diseases mentioned above, it is necessary to continue the search for adequate therapeutics (Setzer, 2013).

For diseases caused by trypanosomatids, such as Chagas disease, African sleeping sickness, and leishmaniasis, exploration of the PA enzyme pathway has been important in drug development (Colotti and Ilari, 2011). PAs are valuable targets for antiparasitic chemotherapy because they play an essential role in the proliferation, differentiation, synthesis of macromolecules and the antioxidant mechanism (Birkholtz et al., 2011; Colotti and Ilari, 2011).

The process of interaction of *T. cruzi* with the host cell involves some important molecules such as Tc85, *trans*-sialidase (TS), and cruzipain (Ouaissi et al., 1986; Alves et al., 1986), which can bind to different components of the host cell, such as cytokeratin 18 (Magdesian et al., 2001), and extracellular matrix components such as fibronectin (Ouaissi et al., 1986) and laminin (Giordano et al., 1994).

7.2.2.1. Tc85

It is an 85-kDa surface glycoprotein belonging to the superfamily gene gp85/TS specific to trypomastigote stage *T. cruzi* (Abuin et al., 1996).

7.2.2.2. Trans*-sialidase*

T. cruzi possesses a membrane-anchored TS enzyme that transfers sialic acids from the host cell surface to the parasitic cell surface, allowing *T. cruzi* to effectively evade the host's immune system. This enzyme has no analogous human counterpart and thus has become an interesting drug target to combat the parasite (Miller and Roitberg, 2013).

7.2.2.3. Cruzipain

Another important *T. cruzi* molecule for the invasion process is cruzipain, an important cysteine proteinase that is able to activate the release of calcium from intracellular stores (de Souza et al., 2010).

The different developmental stages and changing biochemical interactions between the parasite and the vertebrate host during the life cycle of *T. cruzi* render the development of new drugs more difficult (de Souza, 2002). The availability of genomics (El-Sayed et al., 2005) and proteomics data (Atwood et al., 2005) from the parasite has set high hopes for the identification of new drug targets.

Lipid and energy metabolism of the parasite have received considerable attention as potential drug targets in *T. cruzi*.

7.2.2.4. Lipid Metabolism

Lipids have essential roles in biological membranes and are sources of stored energy. They also perform several other functions, such as being cofactors for enzymes, acting as hormones, being intra- and extracellular signal messengers, and being involved in protein anchorage to membranes and transporters. Two enzymes from this pathway are *isopentenyl diphosphate isomerase* and *phophomevalonate kinase*, both of which are analogous to the corresponding human proteins (Alves-Ferreira, unpublished observations).

The steroid metabolism in trypanosomatids was initially studied through the ocalizationion of intermediate metabolites using isotopic ocaliza and inhibitors for several yeast enzymes (Roberts et al., 2003). The presence of variants with substitutions on carbon-24 and the therapeutic effects of azasterols on the parasites clearly show the importance of the *sterol-24-metyltranferase* enzyme in the biosynthesis, besides the overall importance of the ergosteroids in the biology of the parasites. The enzyme sterol-24-metyltransferase has been described in several species of *Leishmania*, in *T. cruzi*, and in *Trypanosoma brucei* (Magaraci et al., 2003; Pourshafie et al., 2004; Jiménez-Jiménez et al., 2006). The exact intracellular localization of the enzyme in *T. cruzi* is still unclear and was proposed to be associated with the glycosome as well as the cytoplasm. In analyses using AnEnPi, three homologous gene copies were identified that encoded this enzyme in *T. cruzi*, corroborating the previous description of this enzyme and its indication as a possible new therapeutic target.

7.2.2.5. The Energy Metabolism

Energy metabolism begins with nutrient uptake. In *T.cruzi*, a single isoform of the hexose transporter THT1 was described and several copies are present in the genome (Bringaud and Baltz, 1993; Tetaud et al., 1994). The transporter in *T. cruzi* has a high affinity for glucose, but can also transport other monosaccharides, such as D-fructose, which differs from the human transporter GLUT1 (Tetaud et al., 1994; Barrett et al., 1998).

7.2.2.5.1. Glycolysis

The enzymes of the glycolytic pathway in trypanosomatids are organized in two cellular compartments, while in higher eukaryotes, these enzymes are present in the cytoplasm. In *T. cruzi*, the first enzyme of the pathway, a *hexokinase* (HK), was described as having different kinetic characteristics than the human enzyme. It did not suffer inhibition by D-glucose-6-phosphate or by other vertebrate HK regulators, such as fructose-1,6-diphosphate, phosphoenolpyruvate, lactate, or citrate, although there is a weak competitive inhibition of ADP with respect to ATP (Racagni and Machado de Domenech, 1983; Urbina and Crespo, 1984). Additional experiments showed that the enzyme is inhibited in a noncompetitive way by inorganic pyrophosphate (PPi) and does not phosphorylate other sugars, such as fructose, mannose, and galactose (Cáceres et al., 2003). This suggests that biphosphonates are possible inhibitors of this *T. cruzi* enzyme. Other reports in this direction showed the potent and selective inhibition of HK and inhibition of the proliferation of the clinically relevant intracellular amastigote form of the parasite, using aromatic arinomethylene biphosphonates, which act as noncompetitive or mixed inhibitors of HK (Hudock et al., 2006; Sanz-Rodriguez et al., 2007).

Cordeiro et al. (2007) showed that the parasite also has a *glucokinase* with 10 times higher affinity for glucose preferentially in a β-D-glucose isomeric form, while the preferential HK substrate is the α-isomer. Using the AnEnPi tool, in silico annotation of *T. cruzi* predicted proteins indicated the presence of both HK and glucokinase (Alves-Ferreira et al., 2009).

Another point of regulation in the glycolytic pathway of eukaryotes resides in the catalytic step of the *phosphofructokinase*. This enzyme displays different characteristics in *T. cruzi*, including the fact that the enzyme is not sensitive to typical modulators, such as the activator fructose-6-P and the inhibitors ATP, citrate, and phosphoenolpyruvate (Urbina and Crespo, 1984; Adroher et al., 1990).

The *triose phosphate isomerase* (TIM) is also a central enzyme of the glycolytic pathway that has been studied in *T. cruzi* and in a number of pathogenic protozoa (Pérez-Montfort et al., 1999; Reyes-Vivas et al., 2001; Rodríguez-Romero et al., 2002; Olivares-Illana et al., 2006). Two studies showed the possibility of developing new therapeutic agents against trypanosomes using TIM as a molecular target. The compound 6,6′-bisbenzothiazole-2,2′ diamine in the low micromolar range was able to specifically inhibit the TIM of trypanosomatids (Olivares-Illana et al., 2006), while the compound dithiodianiline in nanomolar concentrations was able to completely inhibit the recombinant TIM of *T. cruzi* and was trypanocidal for the epimastigote form of the parasite (Olivares-Illana et al., 2007).

The *phosphoglycerate kinase* (PGK) was described by Concepción et al. (2001) as having two isoforms: one of 56 kDa, which is exclusively glycosomal, and a second of 48 kDa, which is expressed both in the cytoplasm and in glycosomes. In the same study, the authors demonstrated that 20% of total PGK activity is found in glycosomes and 80% in the cytoplasm. The *pyruvate kinase* in *T. cruzi* was first reported by Juan et al. (1976), and the following studies demonstrated the kinetic details, regulation, and localization of this enzyme (Cazzulo et al., 1989; Adroher et al., 1990). The enzyme alcohol dehydrogenase (EC 1.1.1.2), which is important in the oxidoreduction of alcohol, can be considered a possible target for therapeutic studies against *T. cruzi* because AnEnPi analysis indicates it as being an analog to the enzyme found in humans (Alves-Ferreira et al., 2009). It was demonstrated that the corresponding enzyme from *Entamoeba histolytica* can be inhibited by cyclopropyl and cyclobutyl carbinols and is considered as a possible new drug target for this parasite as well as for *Giardia lamblia* (Espinosa et al., 2004).

7.2.2.5.2. The Pentose Phosphate Shunt

The pentose phosphate shunt is the most studied pathway in *T. cruzi*, and the participating enzymes have been well characterized both molecularly and biochemically (Igoillo-Esteve et al., 2007). The enzymes of this pathway were identified as cytoplasmic, but with secondary localization in the glycosome. They play an important role in the generation of NADPH, which is essential in responding to oxidative processes of the host defense system. Maugeri and Cazzulo (2004) showed that under normal conditions 10% of the glucose captured by *T. cruzi* is metabolized in the pentose pathway. All seven enzymes of the pathway are expressed in the four life stages of *T. cruzi* (Maugeri and Cazzulo, 2004), and this pathway can be divided into oxidative and nonoxidative branches.

The oxidative branch is responsible for the production of NADPH and ribose-5-phosphate and is regulated by the ratio of NADP/NADPH through the first enzyme in the process, *glucose-6-phosphate dehydrogenase*. This enzyme has several copies in the *T.cruzi* genome, while the *6-phosphogluconolactonase*, the *6-phosphogluconate dehydrogenase*, and the *ribose 5-phosphate isomerase* are present as only one copy. The latter enzyme is analogous to the human enzyme, which makes it an interesting therapeutic target (Igoillo-Esteve et al., 2007).

The nonoxidative branch complements the pathway, and the enzymes involved are *ribulose-5-phosphate epimerase*, with two gene copies in *T. cruzi*, and the *transaldolase* and *transketolase*, each with one copy per haploid genome (Igoillo-Esteve et al., 2007). One of the copies of ribulose-5-phosphate isomerase presents a C-terminal (PTS1) signal for glycosome targeting, but the cytosol also showed activity (Maugeri and Cazzulo, 2004). A transaldolase is responsible for the transfer of the dihydroxyacetone group from fructose-6-phosphate to erythrose-4-phosphate, leading to the synthesis of sedoheptulose-7-phosphate and glyceraldeyde-3-P. The biochemical characterization of the recombinant enzyme was reported (Igoillo-Esteve et al., 2007). The transketolase of *T. cruzi* contains the PTS1 signal peptide, which possibly permits its distribution both in the glycosome and in the cytoplasm. The recombinant enzyme is apparently a dimer of 146 kDa (Igoillo-Esteve et al., 2007).

Using the AnEnPi approach, *sedoheptulose bisphosphatase* was identified (EC 3.1.3.37), which performs the addition of H_2O to sedoheptulose 1,7-bisphosphate, resulting in sedoheptulose 7-phosphate and phosphate. This enzyme is a possible target for the development of drugs because it is specific to the parasite and is absent in the human host (Alves-Ferreira et al., 2009).

7.2.2.5.3. The Krebs Cycle and Oxidative Phosphorylation

The intermediate energy metabolism that occurs in the mitochondria of *T. cruzi* was studied by several groups, as Cannata and Cazzulo (1984) reviewed the metabolism of carbohydrates and described, in detail, current knowledge of the Krebs cycle, as well as the process of aerobic fermentation in this parasite. Within this context, the main products of this peculiar metabolic pathway would be succinate from the reduction of fumarate coupled with the reduction of oxaloacetate and the generation of L-malate. The intermediary energy metabolism was reviewed by Urbina et al. (1993), including a discussion on the possibility of regulating the activity of the Krebs cycle, according to the action of two glutamate dehydrogenases, one NAD^+-dependent and the other $NADP^+$ dependent, which indicates the use of amino acids as the main energy source for the parasite.

The constitutive expression of the *ATP-dependent phosphoenolpyruvate carboxykinase* (PEPCK, EC 4.1.1.49) was described in all evolutionary forms of *T. cruzi*. The enzyme catalyzes the reaction from phosphoenolpyruvate to oxaloacetate yielding ATP. In vertebrates, this enzyme is involved in glycogenesis, while in *T. cruzi* it appears to be involved in catabolism. Additionally, the inhibition of this enzyme by 3-mercaptopicolinic acid has been described in studies with the purified enzyme and in vivo (Urbina et al., 1990). This enzyme was also shown to have a specific requirement for transition metal ions that modulate the reactivity of a single essential thiol group, which differs from the hyperreactive cysteines present in vertebrates or yeast (Jurado et al., 1996). Trapani et al. (2001) elucidated the 3D and dimeric structures of the PEPCK of *T. cruzi*, permitting progress in the study of this enzyme for the development of new therapeutic agents.

Data from in silico analysis indicate the presence of all the enzymes involved in the Krebs cycle, corroborating the analyses by Kyoto Encyclopedia of Genes and Genomes (http://www.genome.jp/kegg/), but also indicated a phosphoenolpyruvate carboxykinase (EC 4.1.1.49) specific for *T. cruzi* and absent in humans. Furthermore, the enzyme fumarate reductase (EC 1.3.99.1) is also specific for *T. cruzi* and fumarate hydratase (EC 4.2.1.2) in the parasite is analogous to the human protein (Alves-Ferreira et al., 2009).

7.2.2.5.4. β-Oxidation

The oxidation of fatty acids is an important source of ATP in many organisms, but this is apparently not the case for most parasites (van Hellemond and Tielens, 2006). Initial analysis of the *T. brucei* genome (Berriman et al., 2005) and also of *T. cruzi* and *Leishmania* identified homologous genes for the four enzymes responsible for β-oxidation of fatty acids, and this pathway probably occurs both in glycosomes and mitochondria (van Hellemond and Tielens, 2006). The oxidation rates, however, seem minimal and oxidation can be for very specific fatty acids only or occurs only under special conditions (Wiemer et al., 1996).

The enzyme *pyruvate phosphate dikinase* was described in *T. cruzi* epimastigotes with a glycosomal localization. The complete function of this enzyme is not completely understood, but its reaction probably leads to pyruvate production from phosphoenolpyruvate and PPi (Acosta et al., 2004). The authors showed that *palmitoyl-CoA β-oxidation* occurs in glycosomes and suggest that this enzyme could be a link between glycolysis, fatty acid oxidation, and the biosynthetic PPi-producing pathways in this organelle, as well as replacing pyrophosphatase in its classical thermodynamic role and eliminating the toxic PPi (Alves-Ferreira et al., 2009).

7.2.2.5.5. Use of Amino Acids for Energy Production

T. cruzi epimastigotes metabolize alanine, aspartate, glutamine, glutamate, leucine, isoleucine, and proline (Sylvester and Krassner, 1976). Alanine, aspartate, and glutamate provide the Krebs cycle with intermediates (Silber et al., 2005), while the latter is an important intermediate in proline metabolism, which is an important carbon and energy source for the parasite. The amino group of glutamate can be transferred to pyruvate by both *alanine aminotransferase* and *tyrosine aminotransferase*, yielding a-ketoglutarate and alanine (Cazzulo, 1994) or can be deaminated by *glutamate dehydrogenase*, which is localized in the cytoplasm and mitochondria (Duschak and Cazzulo, 1991).

A detailed computational analysis of the *T. cruzi* metabolic pathways involving amino acid metabolism were described previously, and many instances of analogous enzymes in the parasite, compared to human, were pointed out (Guimarães et al., 2008). Many questions still remain open regarding energy metabolism in *T. cruzi* and our understanding of the physiological mechanisms involved in the generation of energy in this parasite. Furthermore, the identification of new therapeutic targets needs further work (Alves-Ferreira et al., 2009).

7.2.2.5.6. The Polyamine Metabolism

Due to the importance of PAs for trypanosomatids, which are highly dependent on spermidine for growth and survival, PA metabolism has been well studied, and several enzymes, such as *arginase*, ODC, *S-adenosylmethionine decarboxylase* (AdoMetDC), *spermidine synthase*, *trypanothione synthetase*, and *trypanothione reductase* have been targeted for the development of new drugs (Heby et al., 2007). *T. cruzi* does not contain ODC, an effective target of alpha-difluoromethylornithine for the treatment of sleeping sickness. Instead, the parasite salvages putrescine or spermidine from the host.

Two candidate aminopropyltransferases have been proposed. Although *T. cruzi* maintains an apparently functional copy of the AdoMetDC, known inhibitors of this enzyme do not have much effect on the parasite (Beswick et al., 2006). The traditional cellular redox couple formed by the otherwise *ubiquitous glutathione/glutathione reductase* couple is replaced in trypanosomatids by the dithiol bis(glutathionyl)spermidine called trypanothione and the *flavoenzyme trypanothione reductase*. Trypanothione is the reducing agent of thioredoxin and tryparedoxin, which are small dithiol proteins that are reducing agents for the synthesis of deoxyribonucleotides, as well as for the detoxification of hydroperoxides by different peroxidases. The *trypanothione reductase* is an essential enzyme for the parasite, and its absence in the mammalian host makes it an interesting target for drug development (KrauthSiegel and Inhoff, 2003; Martyn et al., 2007). More detailed analysis of biochemical pathways and increased high-throughput screening activities, using both synthetic compounds and natural products, set high hopes for the development of new drugs against Chagas disease, leishmaniasis, and sleeping sickness. The many ongoing initiatives and the identification of the numerous potential targets bring hope that a breakthrough in the treatment of these parasitic diseases will be shortcoming (Alves-Ferreira et al., 2009).

7.2.3. Human African Trypanosomiasis

One of the most neglected disease is the sleeping sickness or human African trypanosomiasis (HAT), which is mostly restricted to poor regions in Africa. The disease is caused by *T. brucei* after a host has been bitten by the tsetse fly (Scotti et al., 2016). The main drugs used to treat HAT are pentamidine, melarsoprol, suramin, and eflornithine (Steverding, 2010); these are highly toxic and can cause the emergence of resistant forms of the parasite. Also, these drugs are not readily available. Therefore, new drugs are needed urgently (Scotti et al., 2016).

Many researchers are investigating new enzyme targets for the parasite, searching for more efficient and selective inhibitors that are capable of causing parasite death with less toxicity to the host, such as trypanothione reductase, farnesyl diphosphate synthase (FPPS), 6-phosphogluconate dehydrogenase, and UDP 4′-galactose epimerase (Scotti et al., 2016).

7.2.3.1. Trypanothione Reductase

Trypanothione reductase (TR) is essential for parasite survival and is not greatly differentiated between subspecies (Jones et al., 2010; Lu et al., 2013; Zimmermann et al., 2013). The parasite and mammal's glutathione reductase is very similar and protects the parasite from oxidative stress (Patterson et al., 2011; Holloway et al., 2007; Pratt et al., 2014). The parasite's intracellular reduction is maintained by a single thiol antioxidant system. The mammalian reduction of glutathione by glutathione reductase is analogous to trypanothione reduction by TR (Krieger et al., 2000).

7.2.3.2. Farnesyl Diphosphate Synthase

FPPS is a target enzyme studied in infectious, immunological, and bone diseases, among others (Lai et al., 2014). It catalyzes the condensation of the isoprenoid dimethylallyl diphosphate with isopentyl diphosphate (IPP), generating geranyl diphosphate (GPP). This GPP then condenses another IPP molecule into farnesyl diphosphate (FPP) (Scott et al., 2016). Fosamax, Actonel, Zometa, and Aredia are examples of bisphosphonate compounds that are inhibitors of FPPS. Inhibition of other enzymes responsible for the biosynthesis of isoprenes occurs, such as isopentyl diphosphate isomerase, geranyl diphosphate synthase, geranilgeranil diphosphate synthase, 1-deoxixilulose-5-phosphate isomerase, and *T. cruzi* HK (Jones et al., 2010). Using quantum chemistry, spectroscopy, and docking studies; Mao et al. (2006) have investigated a series of bisphosphonates, providing structural information about the FPPS—inhibitor complex against *T. brucei*.

7.2.3.3. N-*myristoyltransferase*

The enzyme NMT is another interesting enzymatic target because it is related to parasite survival. NMT catalyzes the transfer of the myristate group of myristyl-Co-A to the N-terminal glycine, being a constituent of proteins, and influencing the functioning of cell membranes (Scotti et al., 2016).

7.2.3.4. Kinases

The kinases represent a set of enzymes, whose dysfunctions are related to various diseases and represent potential targets. Oduor et al. (2011) performed a virtual screening, selecting inhibitors of glycogen synthase kinase, (anti-glycogen synthase kinase-3 TbruGSK-3). Subsequently, they performed docking in human GSK-3 beta enzyme, analyzing and comparing the interactions for greater selectivity. Of the 40 compounds investigated, two showed greater selectivity by ThuGSK-3.

7.2.3.5. 6-Phosphogluconate Dehydrogenase

Phillips et al. (1998) observed that the enzyme 6-phosphogluconate dehydrogenase from *T. brucei* is only 35% similar to that of other organisms. This makes it extremely favorable as an enzymatic target for structurally selective drug candidates.

7.2.3.6. Uridine Diphosphate Galactose 4′-Epimerase

The enzyme uridine diphosphate galactose 4′-epimerase (UDP-GalE) is oxidoreductase NAD^+-dependent, and catalyzes the interconversion of

UDP-glucose and UDPgalactose (Shaw et al., 2003; Alphey et al., 2006). Many studies have been performed to elucidate the 3D structure of this enzyme, which is critical to parasite survival and has only 33% similarity with the host. The main difference was noted at the active site of the enzyme; in humans, there is a Gly237 residue, and in *T. brucei*, a Cys266 residue (Shaw et al., 2003; Alphey et al., 2006).

7.2.4. Schistosomiasis

Schistosomia mansoni is one of the most widespread human parasites in the world, and the occurrence of schistosomiasis is related to the absence or insufficiency of sanitation. It is endemic in the vast expanse of Brazil and is still considered a serious public health problem because it affects millions of people. The WHO estimated that schistosomiasis affects 200 million people and represents a threat to more than 600 million people living in risk areas belonging to 54 countries, where the parasite is endemic (Ministério da Saúde, 2014).

Schistosomiasis is a parasitic infectious disease caused by flatworms of the genus *Schistosoma*, for which humans have as main etiological agents the species *S. mansoni*, *Schistosoma haematobium*, and *Schistosoma japonicum*, all of them belonging to the class of trematodes of the family Schistosomatidae; they are characterized by the existence of separate sexes, with clear sexual dimorphism. They have as intermediate hosts freshwater snails of the genus *Biomphalaria*, for example, *Biomphalaria glabrata*, which is native to Brazil and has spiral shell and flat shape characteristic of planorbids. The snail measures 3–4 cm in diameter and features a central depression on each side of the shell which is brown (Ministério da Saúde, 2014).

The evolution cycle of this parasite goes through two different stages: the first is the development of the larvae after they penetrate the snail; the second occurs upon leaving the host, and can penetrate humans freely through their skin in the form of cercariae. This penetration occurs in damp places such as streams and ponds. When the parasite takes up residence inside the definitive host, it can be adhered to the liver, gallbladder, or the intestine of humans, causing severe damage, which can evolve from asymptomatic to extremely severe clinical forms and deaths (Ministério da Saúde, 2014).

Schistosomiasis in its various clinical forms is similar to many other diseases. The diagnosis is guided by the patient's history in an endemic area. Confirmation is made by laboratory tests, and additional tests such as stool tests, serologic tests (very useful in areas where disease transmission is low), rectal biopsy, liver biopsy, circulating antigen test for capture ELISA, intradermoreaction, ultrasonography of the abdomen, chest radiography and endoscopy. Treatment consists of a single oral dose of oxamniquine or praziquantel. These drugs are well-tolerated. They exhibit low toxicity, and their efficacy in treatment reaches 80% in adults and 70% in children younger than 15 years. Currently, praziquantel is preferred due to its lower cost (Pordeus et al., 2008).

Only praziquantel is currently available for the treatment and control of schistosomiasis, and the increasing risk of certain strains of schistosomes that are resistant to praziquantel means that the development of new drugs is urgent. With this objective, we have chosen to target the enzymes modifying histones and in particular the histone acetyltransferases and histone deacetylases (HDACs) (Pierce et al., 2011).

7.2.4.1. Histone-Modifying Enzymes as Drug Targets

Histone-modifying enzymes (HME) include HDAC that have been intensively studied as drug targets, but other classes of enzymes, including histone acetyltransferases, *histone methyltransferases*, and *histone demethylases* are increasingly investigated. HME are central actors in the regulation of the epigenetic modification of chromatin, and aberrant epigenetic states often associated with cancer led to interest in HME as targets for therapy. HDACs deacetylate acetylated lysine residues in a variety of proteins, including histones, and also transcription factors and cofactors, and nonnuclear proteins such as tubulin. There has been a considerable effort to develop HDAC inhibitors (HDACi). Dubois et al. (2009) shown that HDACi such as trichostatin A and valproic acid cause the death of *S. mansoni* larvae and adult worms in vitro and that this is probably via the induction of apoptosis in the parasites. Schistosome HDACs, as well as other HME, are therefore promising targets for the development of new drugs (Pierce et al., 2011).

7.2.4.2. Antioxidant Enzymes

Killing of intramolluscan schistosomes by host hemocytes is mediated by reactive oxygen metabolites. Hence, the defense against oxidative damage is essential for the parasite to survive. In the study by Zelcky and Von Janowsky (2004), expression of three key antioxidant enzymes, *superoxide dismutase* (EC 1.15.1.1), *glutathione peroxidase* (EC 1.11.1.9), and *glutathione-S-transferase* (EC 2.5.1.18) was determined in *S. mansoni* miracidia, sporocysts, and cercariae. Stage-dependent expression of these enzymes was shown to be regulated at the transcriptional level. Schistosomes express elevated levels of antioxidant enzymes in interaction with hemocytes from susceptible snail hosts in which they survive. On the other hand, hemocytes of resistant snails may interfere with the reactive oxygen detoxification via down-regulation of schistosome antioxidant enzymes, thus shifting the *balance toward parasite killing*.

7.2.4.3. Gluconeogenic Enzymes

In the study by Tielens et al. (1991), the activities of *glucose-6-phosphatase*, *fructose-1,6-bisphosphatase* (FBPase), PEPCK, and *pyruvate carboxylase* were determined in homogenates of adult *S. mansoni* worms and compared with the activities in homogenates of rat liver and rat skeletal muscle, tissues with a high and low gluconeogenic capacity, respectively. All four gluconeogenic enzymes were present in *S. mansoni*. Experiments with inhibitors of PEPCK gave no indications that this enzyme was involved in the degradation of glucose. This was confirmed by 13C-nuclear magnetic resonance experiments, which indicated that lactate was formed from phosphoenolpyruvate via the actions of pyruvate kinase and lactate dehydrogenase, and that PEPCK did not participate in the formation of lactate. Substrate cycling between fructose-6-dehydrogenase and fructose-1,6-bisphosphate was demonstrated to occur in adult *S. mansoni*. This shows that FBPase participates in the glucose metabolism of this parasite.

7.2.5. Malaria

Malaria is a disease caused by protozoa of the genus Plasmodium, and it is the most prevalent protozoan disease in the world, putting at risk about 40% of world population (about 2.4 billion people) in more than 100 countries (Gomes et al., 2011). The species associated with human malaria are: *Plasmodium falciparum*, *Plasmodium vivax*, *Plasmodium malariae*, and *Plasmodium ovale*. In Brazil, autochthonous transmission of *P. ovale* was never recorded, which is restricted to certain regions of Africa (Ministério da Saúde, 2010a). Malaria is widespread in extensive tropical and subtropical regions, afflicting a great part of the population, especially in developing and underdeveloped nations (Gomes et al., 2011).

The epidemiological situation of malaria in Brazil is worrying today. Although declining, the absolute number of cases in 2008 was still more than 300,000 patients across the country. Of these, 99.9% were transmitted in the legal Amazon, and *P vivax* is the species that causes almost 90% of cases. However, the transmission of *P. falciparum*, known to be responsible for serious and deadly form of the disease, have shown significant reduction in recent years. In addition, the frequency of hospital admissions for malaria in Brazil has also been declining (Ministério da Saúde, 2010a).

Natural malaria transmission occurs through the bite of infected female *Anopheles*, and the most important species is *Anopheles darlingi* whose preferred breeding grounds are collections of clean, warm, and shaded water, and with low flow, very common in the Brazilian Amazon (Ministério da Saúde, 2010a). Infection begins when the parasites (sporozoites) are inoculated into the skin by the bite of the vector, which will invade liver cells (hepatocytes). In these cells, the sporozoites multiply and give rise to thousands of new parasites (merozoites), breaching the hepatocytes and falling into the bloodstream, which invade the red blood cells, beginning the second phase of the cycle, called blood schizogony. It is in this stage that symptoms of malaria appear in the blood (Ministério da Saúde, 2010a).

The clinical manifestations and laboratory findings are quite variable in severe malaria, showing disorders in different organs and organ systems, affecting the central nervous system, and causing severe anemia, renal failure, pulmonary dysfunction, shock, disseminated intravascular coagulation, hypoglycemia, metabolic acidosis, and hepatic dysfunction (Gomes et al., 2011).

Treatment is based on artemisinin derivatives (artesunate and artemether), which are commonly used substances; the first of these two derivatives is available for intravenous, intramuscular, and rectal use, and the second is available only for intramuscular use. These drugs are extracted from *Artemisia annua* (familiy Asteraceae), which is native to China, and they have fast schizontocidal activity, also acting against the gametocytes of *P. vivax* (Gomes et al., 2011).

However, *Plasmodium* has developed resistance to most of the existing drugs, e.g., mefloquine and chloroquine, as well as drug combinations. This is been attributed to the decreased susceptibility of resistance transporters such as PfCRT (chloroquine resistant strains) and PfMDR1 (multidrug resistant strains) (Sen et al., 2007). Therefore, there is a growing need to develop newer drugs and drug combinations to inhibit the growth of protozoa. This requires work on new molecular targets in *Plasmodium spp*., and the design of inhibitors with good parasiticidal activity. In addition, the new drug should be cost-effective to the underprivileged population of the developing world, imposing less financial burden on their governments (Singh and Misra, 2009).

7.2.5.1. Histone Acetyltransferases

Histone acetyltransferases are enzymes that acetylate conserved lysine residues at N-terminal tails of core histone proteins by transferring an acetyl group from acetyl CoA to lysine to form ε-N-acetyl lysine, neutralizing their positive charges. Lysine acetylation and other posttranslational modifications of histones generate binding sites for specific protein–protein interaction domains, such as the acetyl-lysine binding bromodomain (Tang and Eisenbrand, 1992). Its inhibition will lead to transcriptional deactivation, and therefore the enzyme can be used as a novel drug target in *Plasmodium* (Singh and Misra, 2009).

7.2.5.2. Sarcoendoplasmic Reticulum Ca^{2+} ATPase

Sarcoendoplasmic reticulum calcium ATPase (SERCA) is an ATP-coupled Ca^{2+} ion pump involved in metabolic arrest. Ion pumping is one of the most energetically taxing physiological processes in cells, and ion motive ATPases are the likely loci to be differentially regulated in models of metabolic arrest. We propose that the deactivation of SERCA would potentially contribute to the overall suppression of metabolism (Eckstein-Ludwig et al., 2003).

In case of inhibition of activity of these two enzymes *Plasmodium* would not be able to complete its life cycle (Singh and Misra, 2009).

Some studies show that other enzymes may be used as target for treatment of malaria.

7.2.5.3. Histone Deacetylase Inhibitors as Drugs Against Parasite

The use of HDACi against malaria parasites started with the demonstration of the activity of apicidin (a cyclic tetrapeptide HDACi) in inhibiting growth of *P. falciparum* in vitro; it was possible to develop HDACi that are significantly more toxic to the parasite than toward human cells (Andrews et al., 2008; Wheatley et al., 2010).

7.2.5.4. N-myristoyltransferase (ID PDB 2WUU)

NMTs have been well characterized in a range of eukaryotes, including *P. falciparum*, and human cells. They have been shown to be essential for viability in a number of human pathogens, including and *Plasmodium* spp. (Brannigan et al., 2014), being a suitable, potential, and a valid target for antimalaria drug development (Tate et al., 2014; Rackham et al., 2014; Wright et al., 2014; Goncalves et al., 2012).

7.2.5.5. Phosphodiesterase B1

A study by Ogungbe et al. (2014) also found phosphodiesterase B inhibitors. The best compounds were four stilbenoids, which also demonstrated anti-*Plasmodium* activity, and the antiplasmodial flavonoid artonin F (Namdaung et al., 2006).

7.3. FAMILY ASTERACEAE

Asteraceae (Compositeae) is one of the largest families of flowering plants in the world (Hattori and Nakajima, 2008). Some 1600 genera and 24,000 species of this family have been described botanically (Funk et al., 2009), and several revisions regarding its chemistry and biology were published. The two most used are the ratings Bremer and Funk (Bremer, 1996; Funk et al., 2005).

Considered as the family of greatest importance among the phanerogams, representing 10% of the total angiosperm flora (Roque and Bautista, 2008), Asteraceae has a cosmopolitan distribution, being spread across all continents, except Antarctica, but with very wide representation in temperate and semiarid regions of the tropics and subtropics (Roque and Bautista, 2008). The latest classification recognizes 12 subfamilies, and 43 tribes are usually represented by herbaceous plants and small shrubs, rarely by trees (Campos et al., 2016).

The family is very diverse and presents morphology and very complex taxonomy. In 1816, Henry Cassini in his classification system organized the family into 19 tribes, many of which are recognized today. In 1976, the Asteraceae family was divided by Carlquist into two subfamilies: Asteroideae and Cichorioideae, based on morphological studies (Carlquist, 1976). With advances in molecular biology, in the 1980s, molecular studies were carried out in this family, one of the first being reported by Jansen and Palmer (1987), which noted the similarity between the organization of the structure of DNA. In 2005, Funk produced a "supertree" showing the phylogeny of the family Asteraceae, using the most cited studies and unpublished data that were provided by the employees' authors.

Currently, Asteraceae is divided into 12 subfamilies and 43 tribes, and some of the basal clades supported by molecular data are still poorly characterized from a morphological point of view (Funk et al., 2009).

7.4. FLAVONOIDS

In *Chemistry of Natural Products*, secondary metabolites are important chemotaxonomic markers, since they have a restricted distribution and specific botanical sources (Geissman and Crout, 1969; Harborne, 1988). Among them, we highlight the flavonoids, which have the necessary requirements to be successfully used in chemotaxonomy because this class presents great structural diversity. Flavonoids can be found in abundance in the familiy Asteraceae. They are stable, and their chemical structures are relative easy to identify (Stuessy and Bohm, 2001).

Flavonoids are considered one of the largest groups of secondary metabolites of plants and can be found widely in fruits, leaves, teas, and wines. They are important natural pigments in plants, and their main function is to protect these organisms against oxidizing agents (Lopes et al., 2010). Basically, all flavonoids are constituted by three rings, where their carbons may undergo chemical changes such as hydroxylation, hydrogenation, methylation, and sulfonation, leading to the formation of over 4000 flavonoid compounds, which are grouped into classes (Georgiev et al., 2014). They are recommended in daily human diet and are considered by experts of health as important natural protectors of the body against various adverse effects (Ribeiro et al., 2006).

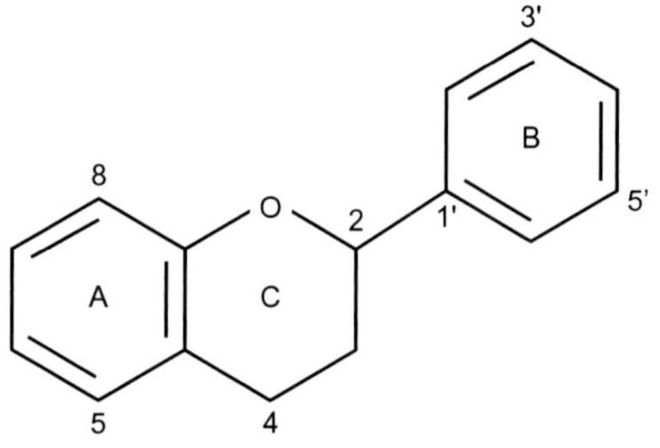

FIGURE 7.2 Basic flavonoid structure.

Flavonoids contain a basic structure that consists of 15 carbon atoms arranged into three rings (C6—C3—C6) and are derived from the shikimate pathway and also via acetate (acetyl coenzyme A). The shikimate pathway affords cinnamic acid and its derivatives (caffeic acid, ferulic, sinapic, etc.) with nine carbon atoms (or C6C3) in the form of coenzyme A, and the route involving acetate gives a tricetídeo to six carbon atoms. Condensation of these derivatives of cinnamic acid with tricetídeo generates a chalcone with 15 carbon atoms, which is the initial precursor of the whole class of flavonoids. From the chalcone, all other flavonoid derivatives are formed (Dewik, 2002).

The chroman system (rings A and C) maintains the second aromatic ring (ring B) on position 2, 3, or 4 (Fig. 7.2). The various types (or skeletons) of flavonoids differ in the degree of oxidation and substitution pattern of the ring C, while the individual substances within a class differ in the substitution pattern of rings A and B. Among the many types of flavonoids, those of particular interest are flavanones, flavones, flavonol, flavan-3-ols, anthocyanidins, isoflavones, chalcones, dihydrochalcones, and aurones due to the variety of substances and diversification of the substitution pattern such as C and O-methyl, O-C- and glycosyl, and O-C- and prenyl (Emerenciano et al., 2007).

This polyphenol class, since the 1980s, stands out for its pharmacological properties (Rodrigues Da Silva et al., 2015). Flavonoids have potential for the treatment of many serious diseases such as neglected tropical diseases, making up drug candidates with antiprotozoal activity.

Several studies have demonstrated the activity of flavonoids on several species of protozoa. According to the literature, 5,6,7-trihydroxy-4′-methoxyflavanone (**1**), an isolated derivative from the MeOH extract of *Baccharis retusa* (Asteaceae), showed activity against cutaneous species of *Leishmania* (Grecco et al., 2010). Fisetin (**2**), quercetin (**3**), luteolin (**4**), and 7,8-dihydroxyflavone (**5**) show high activity in *Leishmania* cultures and present low toxicity to mammalian cells (Manjolin et al., 2013). A set of flavonoides kaempferol (**6**), quercetin (**3**), trifolin (**7**), and acetyl hyperoside (quercetin-3-*O*-β-galactoside acetate, **8**) (Fig. 7.3A) showed leishmanicidal activity against promastigote as well as amastigote forms of *Leishmania spp* (Marín et al., 2009).

Quercitrin (**9**) was demonstrated to kill intracellular amastigotes of *Leishmania amazonensis in vitro* with IC_{50} values of 1 μg/mL (Muzitano et al., 2006a) and 8 μg/mL (18 μM) (Muzitano et al., 2006b). Therefore, the aglycone quercetin (**3**) as well as quercitrin (**9**) and the quercetin-3-*O*-arabinorhamnoside (**10**) (Fig. 7.3B) were tested *in vivo* against *L. amazonensis* in infected BALB/c mice. Intragastric gavage treatment for 30 days with 16 mg flavonoid/kg/day yielded similar efficacy as pentostam (intraperitoneal, 8 mg/kg, twice per week) used as reference drug. Reduction of parasitemia on day 68 ranged from 57% (**9**) up to 76% (**3**) when compared with 62% for pentostam (Muzitano et al., 2009).

Eupafolin (**11**), a 6-methoxyflavone from *Eupatorium perfoliatum* (Asteraceae) was recently shown to possess some antileishmanial (*L. donovani*

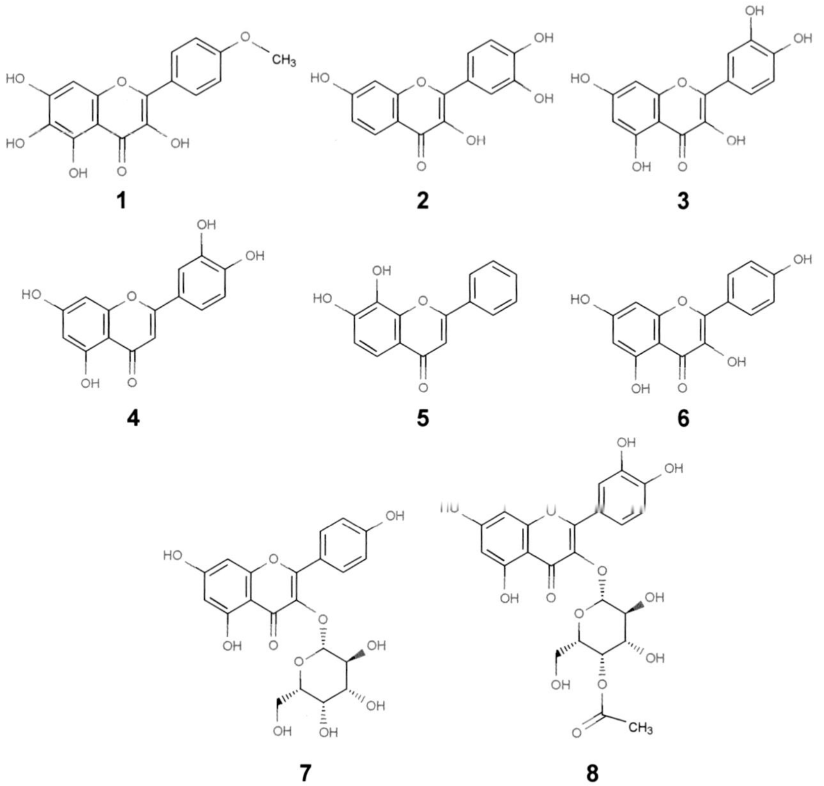

FIGURE 7.3A Structures of flavonoids with antileishmanial activity.

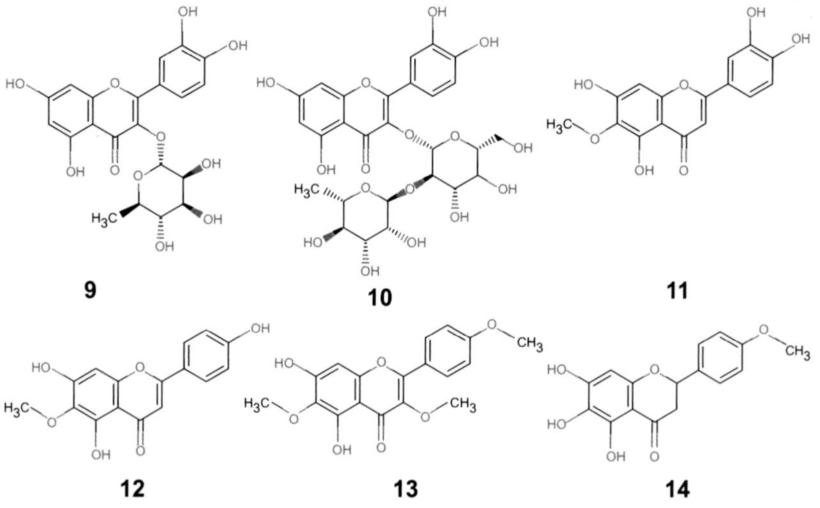

FIGURE 7.3B Structures of flavonoids with antiprotozoal activity.

axenic amastigotes) activity (Maas et al., 2011). Two 6-methoxylated flavonoids, hispidulin (**12**) from *Ambrosia tenuifolia* and santin (**13**) from *Eupatorium buniifolium* (both Asteraceae members growing in Argentina), were found to show *in vitro* activity against *T. cruzi* epimastigotes (IC_{50} = 47 μM) and trypomastigotes (IC_{50} = 62 and 42 μM, respectively) as well as *L. mexicana* promastigotes (IC_{50} = 6 and 32 μM, respetively). The IC_{50}s for cytotoxicity against murine T cells were >50 μM in both cases (Sülsen et al., 2007). A flavanone (**14**) (Fig. 7.3B) isolated from *B. retusa* (Asteraceae) was assayed *in vitro* against promastigotes of various *Leishmania* species (*L. chagasi*, *L amazonensis*, *L. major*, *L. braziliensis*) and showed IC_{50} values between 40 and 54 μg/mL (Grecco et al., 2010).

Besides leishmanicide activity, flavonoids have activity against other types of protozoa as reported in the following studies:

Luteolin (**4**) (Fig. 7.3A) was active against *T. brucei*, *T. cruzi*, *L. donovani*, and *P. falciparum* with IC_{50}s of 3.8, 17.0, 3.0, and 4.2 μg/mL, respectively (Kirmizibekmez et al., 2011). Luteolin (**4**) (Fig. 7.3A) and apigenin (**15**) along with their 7-*O*-glucosides (**16** and **17**, respectively), apigenin-4′-*O*-glucoside (**18**) and rutin **19**, after isolation from *Achillea millefolium* (Asteraceae), were tested against chloroquine-sensitive (D10) and chloroquine-resistant (W2) *P. falciparum* (Fig. 7.3C). Luteolin (**4**) was the most active derivative against both strains (IC_{50} = 6.1 and 5.0 μg/mL, respectively), apigenin (**15**) being

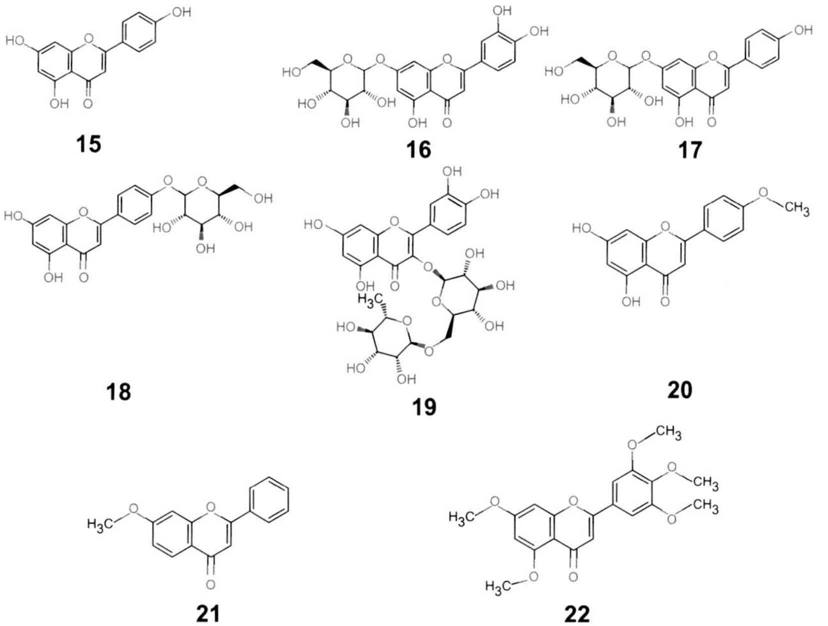

FIGURE 7.3C Structures of flavonoids with antiprotozoal activity.

much less active (IC_{50} = 25.4 and 20.2 μg/mL). However, apigenin-7-*O*-gluoside (**17**) was more active than the corresponding luteolin derivative **16** (IC_{50} = 10.1 and 6.1 μg/mL vs. 26.2 and 26.8 μg/mL). The compound **19** and the 4′-glucoside of apigenin (**18**) were distinctly less active. Several mono- and dicaffeoylquinic acid esters from the same plant did not show any significant activity (Vitalini et al., 2011).

A series of 11 common dietary flavonoids was tested for *in vitro* antiplasmodial activity against *P. falciparum* strains 3D6 (chloroquine sensitive) and 7G8 (chloroquine resistant). The flavone luteolin (**4**) was the most active compound against both strains with IC_{50} values of 11 and 12 μM, respectively, followed by its flavonol congener quercetin (**3**; 15 and 14 μM, respectively). Apigenin (**15**) and acacetin (**20**) (Fig. 7.3C) were also active against the latter strain with IC_{50} = 14 μM. It is remarkable that **20** was selectively active against 7G8 and completely inactive (IC_{50} > 100 μM) against the 3D7 strain while most other compounds affected the two cell lines at similar concentrations (Lehane and Saliba, 2008).

Flavonoids 7-metoxiflavone (**21**) and 3′,4′,5′,5,7-pentametoxiflavone (**22**) (Fig. 7.3C) have shown inhibitory activity against glyceraldehyde 3-phosphate dehydrogenase of *T. cruzi* (Leite et al., 2010), additionally, kaempferol (**6**) promoted the death of adult schistosomes in *in vitro* test at a concentration of 100 μM (Braguine et al., 2012).

Luteolin (**4**), along with some further flavonoids such as the C-glucosylflavone vicenin-2 (**23**), was also tested for activity against *T. cruzi* trypomastigotes after isolation from *Lychnophora pohlii* (Asteraceae). Compounds **4** and **23** (Fig. 7.3D) were the only flavonoids from this study for which IC_{50} values could be determined (1325 and 571 μM, respectively). All other flavonoids did not kill 50% of the parasites at the highest concentration tested (500 μM) (Grael et al., 2005). Three flavanones (eriodictyol **24** and its 3′-mono- as well as 3′,4′-dimethylether, **25** and **26**, respectively), two flavones (luteolin-3′-mono and -7,3′-dimethylether, 27 and **28**), a flavonol (quercetin-7,3′-dimethylether, **29**), and a flavanonol (taxifolin-3′-methylether **30**) (Fig. 7.3D), isolated from *Lychnophora granmongolense* (Asteraceae) along with some sesquiterpene lactones were tested *in vitro* against *T. cruzi* trypomastigotes. Their activity was very low with IC_{50}s ranging from 833 to 2930 μM (Grael et al., 2000).

Six flavones and flavonols (including **3** and several methyl ethers, **31**–**35**) (Fig. 7.3E) from *Chromolaena hirsuta* (Asteraceae) were reported to possess *in vitro* antitrypanosomal activity against *T. cruzi* trypomastigotes. However, the reported IC_{50} values ranging between 102 (**32**) and 352 μg/mL (**33**) were not impressively high. Three of the compounds (**3**, **34**, **35**) were also reported active against *L. amazonensis* promastigotes, compound **35** being the most active with IC_{50} = 87.9 μg/mL (Taleb-Contini et al., 2004).

Methylated flavones from *Lychnophora salicifolia*, namely, quercetin-3,7,3′,4′-tetramethylether (**36**), luteolin-7,3′,4′-trimethylether (**37**), and

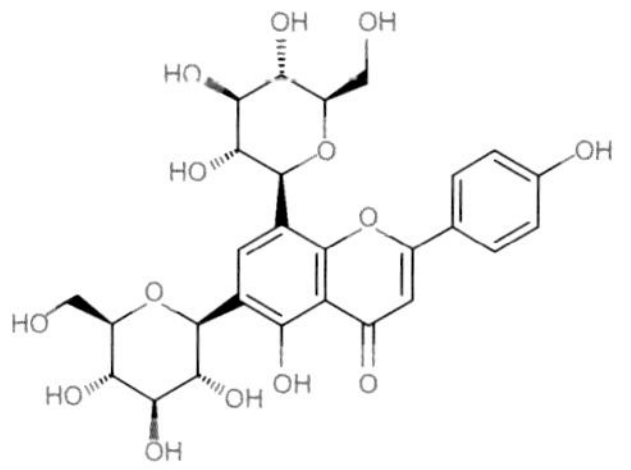

24

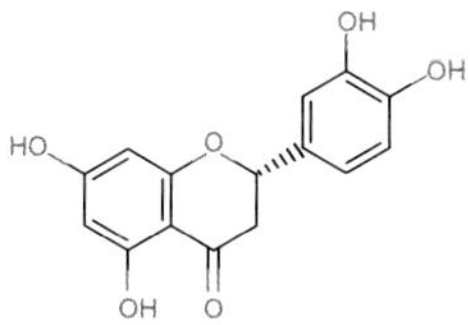

25

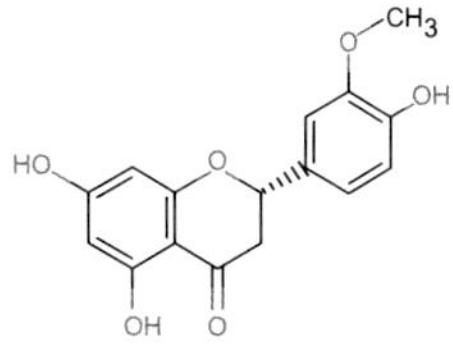

26

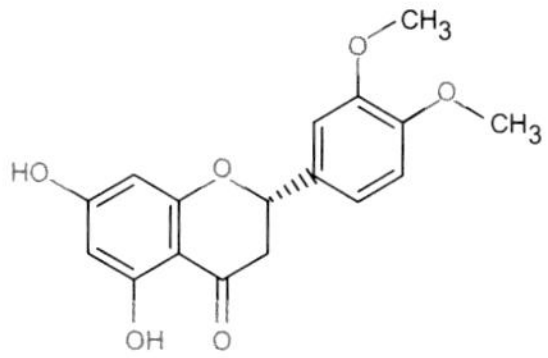

27

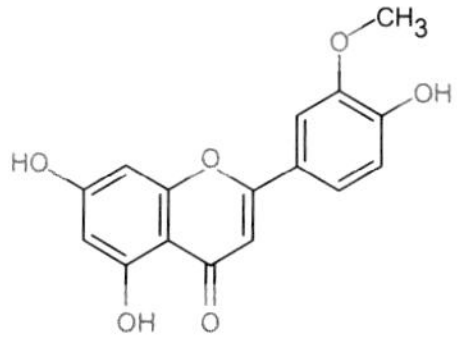

28

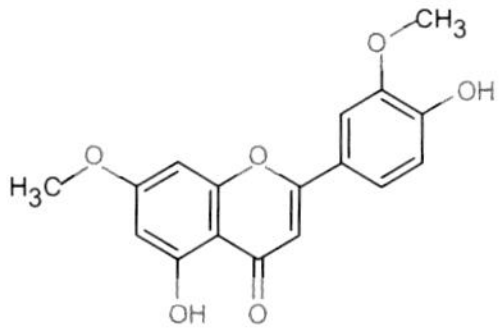

29

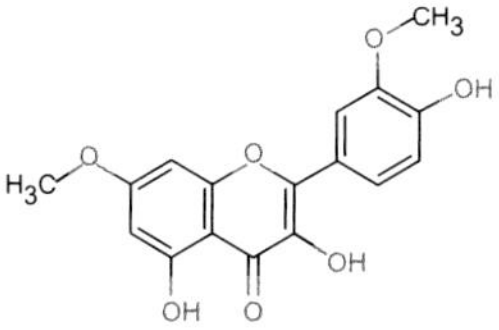

30

FIGURE 7.3D Structures of flavonoids with trypanocidal activity.

quercetin-7,3′,4′-trimethyl ether (**38**), displayed very low *in vitro* activity against *T. cruzi* trypomastigotes ($IC_{50} = 697$, 847 and 217 μM, respectively) (Jordão et al., 2003). Two of four flavonoids isolated from *Trixis vauthieri* (Asteraceae), namely, the flavone penduletin (**39**) and the flavanone sakuranetin (**40**)

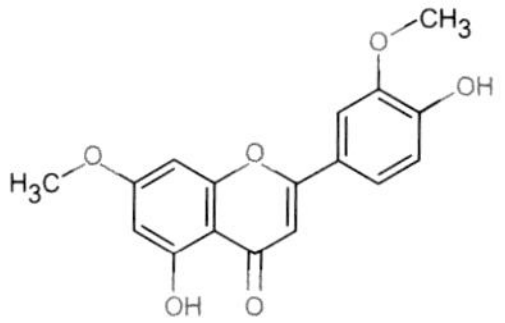

31

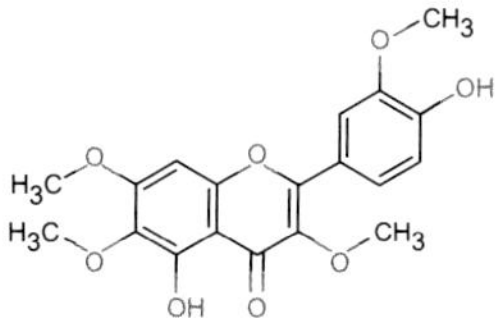

32

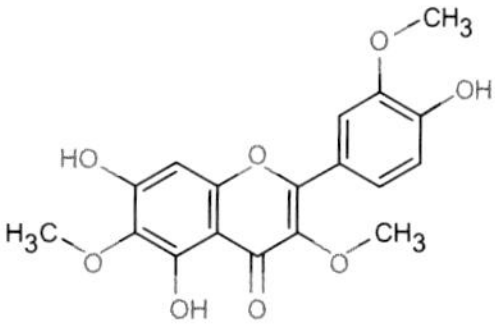

33

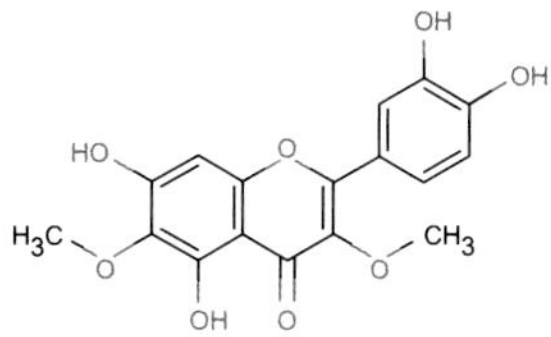

34

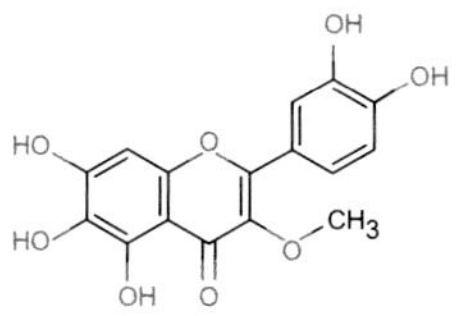

35

FIGURE 7.3E Structures of flavonoids with antiprotozoal activity.

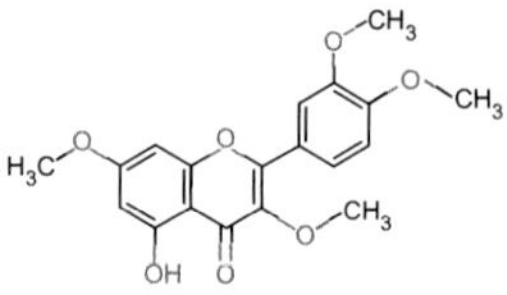
36
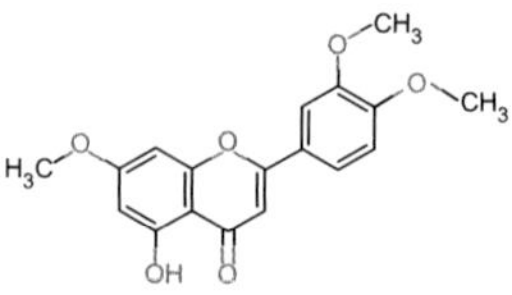
37
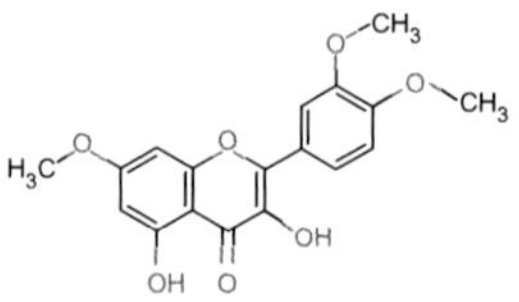
38
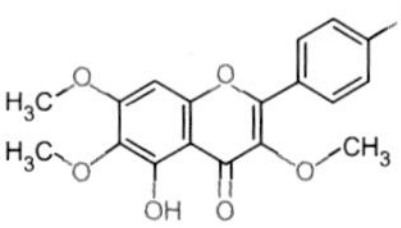
39
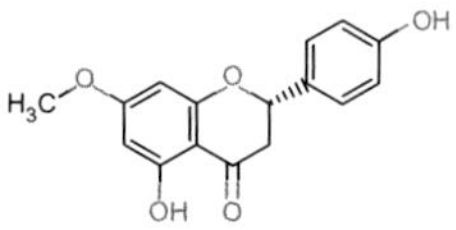
40
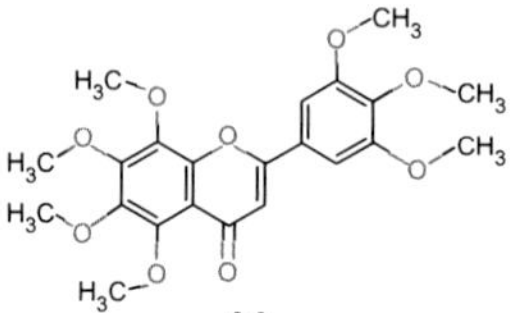
41
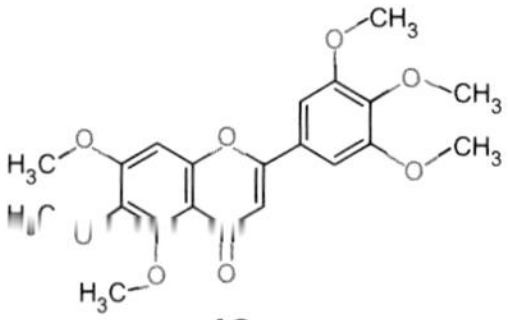
42
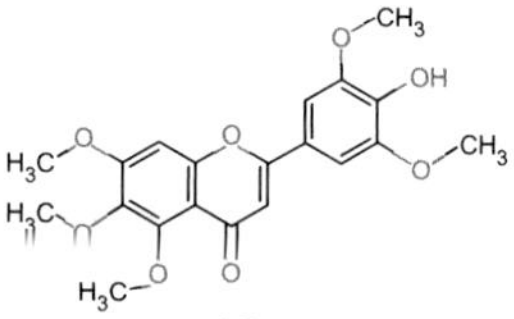
43
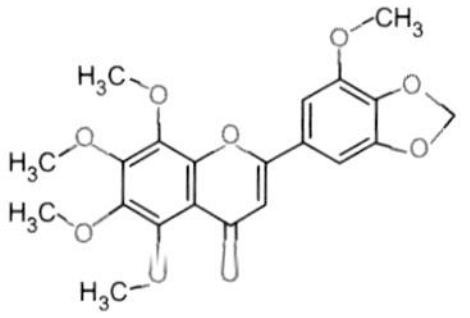
44
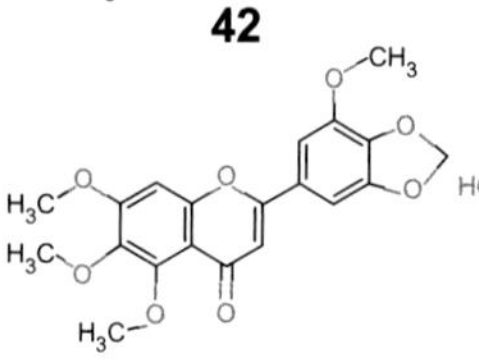
45
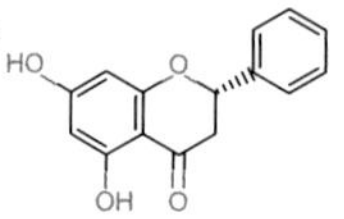
46
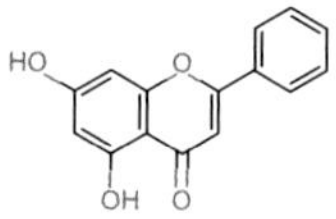
47
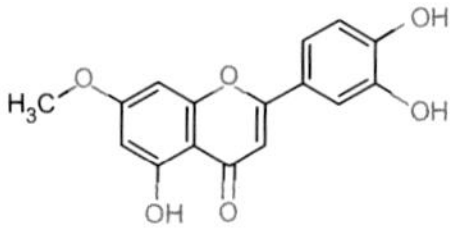
48

FIGURE 7.3F Structures of flavonoids with antiprotozoal activity.

(Fig. 7.3F), displayed some *in vitro* activity against *T. cruzi* trypomastigotes. Both compounds were tested at a concentration of 500 μg/mL and showed 100% and 86% trypanocidal activity at this relatively high concentration (Ribeiro et al., 1997).

Five highly methoxylated/methylenedioxygenated flavones (**41–43**, **44**, **45**) (Fig. 7.3F) were isolated from Sudanese *Ageratum conyzoides* (Asteraceae). While the crude dichloromethane extract of this plant had shown promisingly high *in vitro* activity especially against *T. brucei* ($IC_{50} = 0.78$ μg/mL; SI (selective index) = 47 with L6 cells), none of the constituents isolated in this study (flavonoids and a chromene) showed activity in this range. The most active compound against *T. cruzi* was 4′-hydroxy-5,6,7,3′,5′-pentamethoxyflavone (**43**) with an IC_{50} of 3.0 μg/mL (7.8 μM). This flavone also showed activity in the same concentration range with IC_{50} values of 3.6 μg/mL against *L. donovani* and *P. falciparum* but was inactive against *T. cruzi* ($IC_{50} > 30$ μg/mL) and not cytotoxic against L6 cells ($IC_{50} >$ 90 μg/mL). The compound 5,6,7,3′,4′,5′- hexamethoxyflavone (**42**) was slightly more active against *P. falciparum* (3.0 μg/mL) and also more cytotoxic (SI = 2). Compounds **41** and **45** showed weak activity against *T. cruzi* intracellular amastigotes (IC_{50} = 26.4 and 19.5 μg/mL) (Nour et al., 2010).

Compounds **46**, **47**, and **48** (Fig. 7.3F) showed moderate activity against *Tbr* trypomastigotes (IC_{50} = 10.5, 13.5 and 13.3 μM, respectively) and *P. falciparum* (K1 strain; IC_{50} = 155, 195 and 46 μM) (Camacho et al., 2004).

Probably, the largest set of flavonoids with representatives of all major subtypes (flavones, flavonols, flavanones, flavan-3-ols, isoflavonoids) was investigated by Tasdemir et al. (2006) for *in vitro* activity against *L. donovani* (axenic amastigotes), *T. brucei* (trypomastigotes), and *T. cruzi* (intracellular amastigotes). Cytotoxicity was assessed with L6 cells (SI values); some compounds of other related classes of phenolics were also included. The most active compounds within the respective subgroups were: flavones (32 compounds, including five glycosides and one biflavone): luteolin (**1**) (*L. donovani*: IC_{50} = 0.8 μg/mL; SI = 12), 7,8-dihydroxyflavone (**49**) (*T. brucei*: IC_{50} = 0.068 μg/mL; SI = 116), and chrysin dimethylether (**50**) (*T. cruzi*:

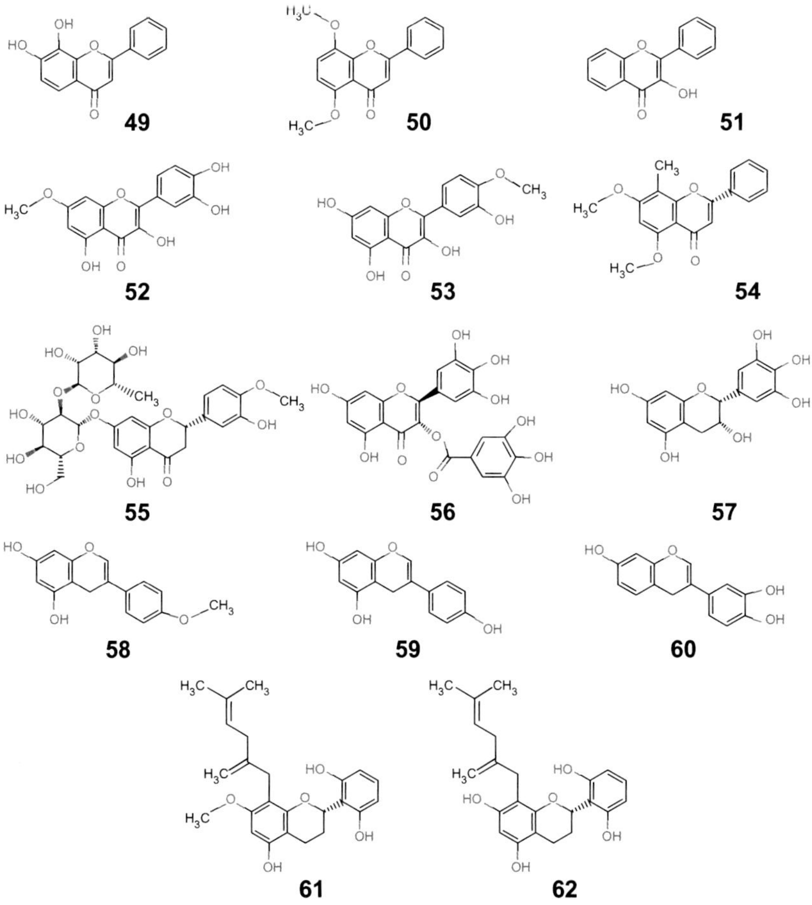

FIGURE 7.3G Structures of flavonoids with antiprotozoal activity.

$IC_{50} = 3.9$ μg/mL; SI = 6); flavonols (25 compounds including five glycosides): fisetin (**2**) (*L. donovani*: $IC_{50} = 0.7$ μg/mL; SI = 64), 3-hydroxyflavone and rhamnetin (**51**,**52**) (*T. brucei*: both $IC_{50} = 0.5$ μg/mL; SI = 21, >180, respectively), and tamarixetin (**53**) (*T. cruzi*: $IC_{50} = 6.4$ μg/mL; SI = 7); flavanones (7 compounds, including two glycosides): 5,7-dimethoxy-8-methylflavanone (**54**) (*L. donovani*: $IC_{50} = 2.4$ μg/mL and *T. cruzi*: $IC_{50} = 13.6$ μg/mL; SI = 16 and 3, respectively) and neohesperidin (**55**) (*T. brucei*: $IC_{50} = 11.5$ μg/mL; SI > 7.8); flavan-3-ols (catechins; 10 compounds): (−)-gallocatechin gallate (**56**) (*L. donovani*: $IC_{50} = 8.9$ μg/mL; *T. brucei*: $IC_{50} = 3.7$ μg/mL; SI = 1.7 and 4, respectively) and (−)-epigallocatechin (**57**) (*T. cruzi*: $IC_{50} = 80.7$ μg/mL; SI = 0.2); isoflavones: biochanin A (**58**) (*L. donovani*: $IC_{50} = 2.5$ μg/mL; SI = 26), genistein (**59**) (*T. brucei*: $IC_{50} = 1.3$ μg/mL; SI = 16), and 3′-hydroxydaidzein (**60**) (Fig. 7.3G) (*T. cruzi*: $IC_{50} = 4.7$ μg/mL; SI = 4.5).

Two lavandulyl flavanones, exiguaflavanones A and B (**61**, **62**) (Fig. 7.3G), structurally related to those from *Sophora flavescens*, were obtained from *Artemisia indica* (Asteraceae). The *in vitro* activity against *P. falciparum* (K1 strain) had been somewhat lower than the ones just mentioned, with IC_{50}s of 4.6 and 7.1 μg/mL, respectively (Chanphen et al., 1998).

In studies involving flavonoid compounds such as enzymatic inhibitors in *Leishmania*, arginase is the most investigated enzyme. An interesting study was conducted by Da Silva et al. (2012) with quercetin (**3**) and quercitrin (9) as arginase inhibitors. Similarly, Manjolin et al. (2013) studied flavonoids (as arginase inhibitors) and subsequently, these same compounds were subjected to docking. The authors observed that fisetin (**2**) was four times more potent than quercetin (**3**), indicating that the hydroxyl at position 5 may not be necessary to inhibit arginase. Yet, quercetin (**3**), which has a hydroxyl at position 3, is twice as potent as luteolin.

In these studies, with flavonoids as arginase inhibitors, their leishmanicidal activities confirm these compounds as new lead candidates in the search for leishmanicidal drugs.

Pleiotropic drugs against cancer as well as cardiovascular and parasitic diseases have been considered by several researchers (Cavalli et al., 2010; Frantz, 2005; Hampton, 2004). Considered as multitarget compounds against *Leishmania* and *Trypanosoma* (Cavalli et al., 2010), the flavonoids can comprise multifunctional drugs or can be used as lead compounds in multifunctional drug design schemes. The biodiversity of flavonoids can contribute to research based on the mechanism of action of isolated or mixed flavonoids on multiple targets related to leishmaniasis (Cruz et al., 2013) and also other most threatening protozoan diseases.

7.5. CONCLUSION

Billions of people are affected by neglected tropical diseases. The drugs currently in use are lacking efficacy, are toxic, and show other liabilities such as parenteral application or high cost. Drug discovery for neglected diseases is not widely met with the necessary attention and determination that the many patients and people at risk deserve.

Clearly, there is an urgent need to discover new chemical structures and novel mechanisms of action to overcome these problems. The search for potentially new leads against the major pathogens responsible for some of the most threatening protozoan diseases, namely, species of *Leishmania*, *Trypanosoma*, *Plasmodium*, and *Schistosoma*, showed the antiprotozoal activity of flavonoids from the family Asteraceae. This confirms that these compounds may be new drug candidates against protozoal diseases due to their multitarget activity.

REFERENCES

Abuin, G., Couto, A.S., de Lederkremer, R.M., Casal, O.L., Gall, I.C., Colli, W., Alves, M.J., 1996. *Trypanosoma cruzi*: the Tc-85 surface glycoprotein shed by trypomastigotes bears a modified glycosylphosphatidylinositol anchor. Exp. Parasitol. 82 (3), 290–297.

Acosta, H., Dubourdieu, M., Quiñones, W., Cáceres, A., Bringaud, F., Concepción, J.L., 2004. Pyruvate phosphate dikinase and pyrophosphate metabolism in the glycosome of *Trypanosoma cruzi* epimastigotes. Comp. Biochem. Physiol. B 138, 347–356.

Adroher, F.J., Osuna, A., Lupiáñez, J.Á., 1990. Differential energetic metabolism during *Trypanosoma cruzi* differentiation. II. Hexokinase, phosphofructokinase and pyruvate kinase. Mol. Cell. Biochem. 94, 71–82.

Agência Brasil, 2015. OMS pede investimentos no combate a doenças tropicais negligenciadas. Available at: http://agenciabrasil.ebc.com.br/internacional/noticia/2015-02/oms-pede-investimentos-no-combate-doencas-tropicais-negligenciadas.

Alves, M.J., Abuin, G., Kuwajima, V.Y., Colli, W., 1986. Partial inhibition of trypomastigote entry into cultured mammalian cells by monoclonal antibodies against a surface glycoprotein of *Trypanosoma cruzi*. Mol. Biochem. Parasitol. 21, 75–82.

Alves-Ferreira, M., Guimarães, A.C., Capriles, P.V., Dardenne, L.E., Degrave, W.M., 2009. A new approach for potential drug target discovery through in silico metabolic pathway analysis using *Trypanosoma cruzi* genome information. Mem. Inst. Oswaldo Cruz 104 (8), 1100–1110.

Alphey, M.S., Burton, A., Urbaniak, M.D., Boons, G.J., Ferguson, M.A.J., Hunter, W.N., 2006. *Trypanosoma brucei* UDP-galactose-4′-epimerase in ternary complex with NAD(+) and the substrate analogue UDP-4-deoxy-4-fluoro-alpha-D-galactose. Acta Crystallogr. Sect. F Struct. Biol. Cryst. Commun. 62, 829–834.

Andrade, B.L.A., Rocha, D.G., 2015. Doenças negligenciadas e bioética: diálogo de um velho problema com uma nova área do conhecimento. Rev. Bioét. 23 (1), 105–113.

Andrews, K.T., Tran, T.N., Lucke, A.J., Kahnberg, P., Le, G.T., Boyle, G.M., Gardiner, D.L., Skinner-Adams, T.S., Fairlie, D.P., 2008. Potent antimalarial activity of histone deacetylase inhibitor analogues. Antimicrob. Agents Chemother. 52, 1454–1461.

Anvisa, 2008. Gerenciamento do Risco Sanitário na Transmissão de Doença de Chagas Aguda por Alimentos, pp. 1–9.

Asilian, A., Sadeghinia, A., Faghihi, G., Momeni, A., 2004. Int. J. Dermatol. 43, 281.

Assad, L., 2010. Doenças negligenciadas estão nos países pobres e em desenvolvimento. Ciênc. Cult. 62 (1), 6–8.

Atwood 3rd, J.A., Weatherly, D.B., Minning, T.A., Bundy, B., Cavola, C., Opperdoes, F.R., Orlando, R., Tarleton, R.L., 2005. The *Trypanosoma cruzi* proteome. Science 309, 473–476.

Barrett, M.P., Tetaud, E., Seyfang, A., Bringaud, F., Baltz, T., 1998. Trypanosome glucose transporters. Mol. Biochem. Parasitol. 91, 195–205.

Barros-Alvarez, X., Gualdrón-López, M., Acosta, H., Cáceres, A.J., Graminha, M.A.S., Michels, P.A.M., Concepción, J.L., Quiñones, W., 2014. Glycosomal targets for anti-trypanosomatid drug discovery. Curr. Med. Chem. 21 (15), 1679–1706.

Bastos, M.M., Boechat, N., Gomes, A.T.P.C., Neves, M.G.P.M.S., Cavaleiro, J.A.S., 2012. O Uso de Porfirinas em Terapia Fotodinâmica no Tratamento da Leishmaniose Cutânea. Rev. Virtual Quím. 4, 108–119.

Berriman, M., Ghedin, E., Hertz-Fowler, C., Blandin, G., Renauld, H., Bartholomeu, D.C., Lennard, N.J., Caler, E., Hamlin, N.E., Haas, B., Böhme, U., Hannick, L., Aslett, M.A., Shallom, J., Marcello, L., Hou, L., Wickstead, B., Alsmark, U.C.M., Arrowsmith, C., Atkin, R.J., Barron, A.J., Bringaud, F., Brooks, K., Carrington, M., Cherevach, I., Chillingworth, T.-J., [illegible], Clark, L.N., Corton, C.H., Cronin, A., Davies, R.M., Doggett, J., Djikeng, A., Feldblyum, T., Field, M.C., Fraser, A., Goodhead, I., [illegible], Harper, D., Harris, B.R., Hauser, H., Hostetler, J., Ivens, A., Jagels, K., Johnson, D., Johnson, J., Jones, K., Kerhornou, A.X., Koo, H., Larke, N., Landfear, S., Larkin, C., Leech, V., Line, A., Lord, A., MacLeod, A., Mooney, P.J., Moule, S., Martin, D.M.A., Morgan, G.W., Mungall, K., Norbertczak, H., Ormond, D., Pai, G., Peacock, C.S., Peterson, J., Quail, M.A., Rabbinowitsch, E., Rajandream, M.-A., Reitter, C., Salzberg, S.L., Sanders, M., Schobel, S., Sharp, S., Simmonds, M., Simpson, A.J., Tallon, L., Turner, C.M.R., Tait, A., Tivey, A.R., Van Aken, S., Walker, D., Wanless, D., Wang, S., White, B., White, O., Whitehead, S., Woodward, J., Wortman, J., Adams, M.D., Embley, T.M., Gull, K., Ullu, E., Barry, J.D., Fairlamb, A.H., Opperdoes, F., Barrell, B.G., Donelson, J.E., Hall, N., Fraser, C.M., Melville, S.E., El-Sayed, N.M., 2005. The genome of the African trypanosome *Trypanosoma brucei*. Science 309, 416–422.

Beswick, T.C., Willert, E.K., Phillips, M.A., 2006. Mechanisms of allosteric regulation of *Trypanosoma cruzi* S-adenosylmethionine decarboxylase. Biochemistry 45, 7797–7807.

Birkholtz, L.M., Williams, M., Niemand, J., Louw, A.I., Persson, L., Heby, O., 2011. Polyamine homoeostasis as a drug target in pathogenic protozoa: peculiarities and possibilities. Biochem. J. 438 (2), 229–244.

Boitz, J.M., Ullman, B., 2013. Adenine and adenosine salvage in *Leishmania donovani*. Mol. Biochem. Parasitol. 190 (2), 51–55.

Boitz, J.M., Strasser, R., Hartman, C.H., Jardim, A., Ullman, B., 2012. Adenine amino hydrolase from *Leishmania donovani*. J. Biol. Chem. 287 (10), 7626–7639.

Braguine, C.G., Bertanha, C.S., Gonçalves, U.O., Magalhães, L.G., Rodrigues, V., Melleiro Gimenez, V.M., Groppo, M., Silva, M.L., Cunha, W.R., Januário, A.H., Pauletti, P.M., 2012. Schistosomicidal evaluation of flavonoids from two species of Styrax against *Schistosoma mansoni* adult worms. Pharm. Biol. 50 (7), 925–929.

Brannigan, J.A., Roberts, S.M., Bell, A.S., Hutton, J.A., Hodgkinson, M.R., Tate, E.W., Leatherbarrow, R.J., Smith, D.F., Wilkinsona, A.J., 2014. Diverse modes of binding in structures of *Leishmania major* N-myristoyltransferase with selective inhibitors. IUCrJ 1 (4), 250–260.

Brannigan, J.A., Smith, B.A., Yu, Z., Brzozowski, A.M., Hodgkinson, M.R., Maroof, A., Price, H.P., Meier, F., Leatherbarrow, R.J., Tate, E.W., Smith, D.F., Wilkinson, A.J., 2010. *N*-myristoyltransferase from *Leishmania donovani*: structural and functional characterisation of a potential drug target for visceral leishmaniasis. J. Mol. Biol. 396 (4), 985–999.

Bremer, K., 1996. Compositae: Systematics. Proceedings of the International Compositae Conference. Royal Botanic Garden, Kew. 1, 1–7.

Bringaud, F., Baltz, T., 1993. Differential regulation of two distinct families of glucose transporter genes in *Trypanosoma brucei*. Mol. Cell. Biol. 13, 1146–1154.

Cáceres, A.J., Portillo, R., Acosta, H., Rosales, D., Quinõnes, W., Avilan, L., Salazar, L., Dubourdieu, M., Michels, P.A.M., Concepción, J.L., 2003. Molecular and biochemical characterization of hexokinase from *Trypanosoma cruzi*. Mol. Biochem. Parasitol. 126, 251–262.

Camacho, M.R., Phillipson, J.D., Croft, S.L., Yardley, V., Solis, P.N., 2004. In vitro antiprotozoal and cytotoxic activities of some alkaloids, quinones, flavonoids, and coumarins. Planta Méd. 70, 70–72.

Campos, F.R., Bressan, J., Jasinski, V.C.G., Zuccolotto, T., Silva, L.E., Cerqueira, L.B., 2016. *Baccharis* (Asteraceae): chemical constituents and biological activities. Chem. Biodivers. 13, 1–17.

Cannata, J.J., Cazzulo, J.J., 1984. The aerobic fermentation of glucose by *Trypanosoma cruzi*. Comp. Biochem. Physiol. B 79, 297–308.

Carlquist, S., 1976. Tribal interrelationships and phylogeny of the Asteraceae. Aliso 8, 465–492.

Castro, J.A., 2000. Contribution of toxicology to the problem of Chagas disease (American trypanosomiasis). Biomed. Environ. Sci. 13, 271–279.

Castro, J.A., Meca, M.M., Bartel, L.C., 2006. Toxic side effects of drugs used to treat Chaga's disease (American trypanosomiasis). Hum. Exp. Toxicol. 25, 471–479.

Cavalli, A., Lizzi, F., Bongarzone, S., Belluti, F., Piazzi, L., Bolognesi, M.L., 2010. Complementary medicinal chemistry-driven strategies toward new antitrypanosomal and antileishmanial lead drug candidates. FEMS Immunol. Med. Microbiol. 58, 51–60.

Cavazzuti, A., Pagliett, G., Hunter, W.N., Gamarro, F., Piras, S., Loriga, M., Alleca, S., Corona, P., McLuskey, K., Tulloch, L., Gibellini, F., Ferrari, S., Costi, M.P., 2008. Discovery of potent pteridine reductase inhibitors to guide antiparasite drug development. Proc. Natl. Acad. Sci. 105 (5), 1448–1453.

Cazzulo, J.J., 1994. Intermediate metabolism in *Trypanosoma cruzi*. J Bioenerg. Biomembr. 26, 157–165.

Cazzulo, J.J., Cazzulo Franke, M.C., Franke de Cazzulo, B.M., 1989. On the regulatory properties of the pyruvate kinase from *Trypanosoma cruzi* epimastigotes. FEMS Microbiol. Lett. 50, 259–263.

Center for Disease Control and Prevention, 2016. Neglected Tropical Diseases. Available at: http://www.cdc.gov/globalhealth/ntd/fastfacts.html.

Chanphen, R., Thebtaranonth, Y., Wanauppathamku, S., Yuthavong, Y., 1998. Antimalarial principles from *Artemisia indica*. J. Nat. Prod. 61, 1146–1147.

Chawla, B., Madhubala, R., 2010. Drug targets in *Leishmania*. J. Parasit. Dis. 34 (1), 1–13.

Colotti, G., Ilari, A., 2011. Polyamine metabolism in *Leishmania*: from arginine to trypanothione. Amino Acids 40 (2), 269–285.

Concepción, J.L., Adjé, C.A., Quiñones, W., Chevalier, N., Dubourdieu, M., Michels, P.A., 2001. The expression and intracellular distribution of phosphoglycerate kinase isoenzymes in *Trypanosoma cruzi*. Mol. Biochem. Parasitol. 118, 111–121.

Cordeiro, A.T., Cáceres, A.J., Vertommen, D., Concepción, J.L., Michels, P.A., Versées, W., 2007. The crystal structure of *Trypanosoma cruzi* glucokinase reveals features determining oligomerization and anomer specificity of hexose-phosphorylating enzymes. J. Mol. Biol. 372, 1215–1226.

Cruz Ede, M., da Silva, E.R., Maquiaveli Cdo, C., Alves, E.S., Lucon Jr., J.F., dos Reis, M.B., de Toledo, C.E., Cruz, F.G., Vannier-Santos, M.A., 2013. Leishmanicidal activity of *Cecropia pachystachya* flavonoids: arginase inhibition and altered mitochondrial DNA arrangement. Phytochemistry 89, 71–77.

Correa, R., Laciar, E., Arini, P., Jane, R., 2010. Analysis of QRS loop in the Vectorcardiogram of patients with Chagas' disease. Conf. IEEE Eng. Med. Biol. Soc. 1, 2561–2564.

Da Silva, M.F.L., Zampieri, R.A., Muxel, S.M., Beverley, S.M., Floeter-Winter, L.M., 2012. *Leishmania amazonensis* arginase compartmentalization in the glycosome is important for parasite infectivity. PLoS One 7 (3), e34022.

de Souza, W., 2002. From the cell biology to the development of new chemotherapeutic approaches against trypanosomatids: dreams and reality. Kinetoplastid Biol. Dis. 1, 3.

de Souza, W., Deli, T.M., Barrias, E.S., 2010. Review on *Trypanosoma cruzi:* host cell interaction. Cell. Biol. Int. 1–18.

Dewik, P.M., 2002. Medicinal Natural Products. A Biosynthetic Approach, second ed. John Wiley & Sons, Chichester.

Dubois, F., Caby, S., Oger, F., Cosseau, C., Capron, M., Grunau, C., Dissous, C., Pierce, R.J., 2009. Histone deacetylase inhibitors induce apoptosis, histone hyperacetylation and up-regulation of gene transcription in *Schistosoma mansoni*. Mol. Biochem. Parasitol. 168, 7–15.

Duschak, V.G., Cazzulo, J.J., 1991. Subcellular localization of glutamate dehydrogenases and alanine aminotransferase in epimastigotes of *Trypanosoma cruzi*. FEMS Microbiol. Lett. 67, 131–135.

Eckstein-Ludwig, U., et al., 2003. Nature 424, 957.

El-Sayed, N.M., Myler, P.J., Bartholomeu, D.C., Nilsson, D., Aggarwal, G., Tran, A.N., Ghedin, E., Worthey, E.A., Delcher, A.L., Blandin, G., Westenberger, S.J., Caler, E., Cerqueira, G.C., Branche, C., Haas, B., Anupama, A., Arner, E., Aslund, L., Attipoe, P., Bontempi, E., Bringaud, F., Burton, P., Cadag, E., Campbell, D.A., Carrington, M., Crabtree, J., Darban, H., da Silveira, J.F., de Jong, P., Edwards, K., Englund, P.T., Fazelina, G., Feldblyum, T., Ferella, M., Frasch, A.C., Gull, K., Horn, D., Hou, L., Huang, Y., Kindlund, E., Klingbeil, M., Kluge, S., Koo, H., Lacerda, D., Levin, M.J., Lorenzi, H., Louie, T., Machado, C.R., McCulloch, R., McKenna, A., Mizuno, Y., Mottram, J.C., Nelson, S., Ochaya, S., Osoegawa, K., Pai, G., Parsons, M., Pentony, M., Pettersson, U., Pop, M., Ramirez, J.L., Rinta, J., Robertson, L., Salzberg, S.L., Sanchez, D.O., Seyler, A., Sharma, R., Shetty, J., Simpson, A.J., Sisk, E., Tammi, M.T., Tarleton, R., Teixeira, S., Van Aken, S., Vogt, C., Ward, P.N., Wickstead, B., Wortman, J., White, O., Fraser, C.M., Stuart, K.D., Andersson, B., 2005. The genome sequence of *Trypanosoma cruzi*, etiologic agent of Chagas disease. Science 309, 409–415.

Emerenciano, V.P., Barbosa, K.O., Scotti, M.T., Ferreira, M.J.P., 2007. Self-organizing maps in chemotaxonomic studies of Asteraceae: a classification of tribes using flavonoid data. J. Braz. Chem. Soc. 18 (5), 891–899.

Espinosa, A., Clark, D., Stanley Jr., S.L., 2004. *Entamoeba histolytica* alcohol dehydrogenase 2 (EhADH2) as a target for anti-amoebic agents. J. Antimicrob. Chemother. 54, 56–59.

Equipe Médica do MSF, 2012. O assunto é doenças negligenciadas. Available at: http://www.msf.org.br/noticias/o-assunto-e-doencas-negligenciadas.

Filho, V.C., Yunes, R.A., 1998. Estratégias para a Obtenção de Compostos Farmacologicamente Ativos a partir de Plantas Medicinais. Conceitos sobre Modificação Estrutural para Otimização da Atividade. Quím. Nova 21, 99–105.

Fiocruz, 2012. Novos casos de Doença de Chagas no Brasil se concentram no Pará e Amapá. Available at: http://www.fiocruz.br/pidc/cgi/cgilua.exe/sys/start.htm?infoid=144&sid=20.

Frantz, S., 2005. Drug discovery: playing dirty. Nature 437, 942–943.

French, J.B., Yates, P.A., Soysa, D.R., Boitz, J.M., Carter, N.S., Chang, B., Ullman, B., Ealick, S.E., 2011. The *Leishmania donovani* UMP synthase is essential for promastigote viability and has an unusual tetrameric structure that exhibits substrate-controlled oligomerization. J. Biol. Chem. 286 (23), 20930–20941.

Funk, V., Bayer, R.J., Keeley, S., Chan, R., Watson, L., Gemeinholzer, B., Schilling, E., Panero, J.L., Baldwin, B.G., Garcia-Jacas, N., Susanna, A., Jansen, R.K., 2005. Everywhere but Antarctica: using a supertree to understand the diversity and distribution of the Compositae. Biol. Skr. 55, 343–374.

Funk, V.A., Susanna, A., Stuessy, T.F., Robinson, H., 2009. Classification of compositae. In: Funk, V.A., Susanna, A., Stuessy, T.F., Bayer, R.J. (Eds.), Systematics, Evolution, and Biogeography of Compositae. International Association for Plant Taxonomy. Vienna, Austria, pp. 171–189.

Gasteiger, J. (Ed.), 2003. Handbook of Chemoinformatics: From Data to Knowledge in 4 Volumes. Wiley-VCH Verlag GmbH, Weinheim.

Gasteiger, J., Engel, T. (Eds.), 2003. Chemoinformatics: A Textbook. Wiley-VCH Verlag GmbH & Co. KGaA, Weinheim.

Geissman, T.A., Crout, D.H.G., 1969. Organic Chemistry of Secondary Plant Metabolism. Freeman Cooper & Company, San Francisco, CA, USA.

Georgiev, V., Ananga, A., Tsolova, V., 2014. Recent advances and uses of grape flavonoids as nutraceuticals. Nutrients 6 (1), 391–415.

Gil, E.S., Paula, J.R., Nascimento, F.R.F., Bezerra, J.C.B., 2008. Produtos naturais com potencial leishmanicida. Rev. Ciênc. Farm. Básica Apl. 29 (3), 223–230.

Giordano, R., Chammas, R., Veiga, S.S., Colli, W., Alves, M.J.M., 1994. An acidic component of the heterogeneous Tc-85 protein family from the surface of *Trypanosoma cruzi* is a laminin binding glycoprotein. Mol. Biochem. Parasitol. 65, 85–94.

Goldston, A.M., Sharma, A.I., Paul, K.S., Engman, D.M., 2014. Acylation in trypanosomatids: an essential process and potential drug target. Trends Parasitol. 30 (7), 350–360.

Gomes, A.P., Vitorino, R.R., Costa, A.P., Mendonça, E.G., Oliveira, M.G.A., Batista, R.S., 2011. Malária grave por *Plasmodium falciparum*. Rev. Bras. Ter. Intensiv. 23 (3), 358–369.

Goncalves, V., Brannigan, J.A., Whalley, D., Ansell, K.H., Saxty, B., Holder, A.A., Wilkinson, A.J., Tate, E.W., Leatherbarrow, R.J., 2012. Discovery of Plasmodium vivax *N*-myristoyltransferase inhibitors: screening, synthesis, and structural characterization of their binding mode. J. Med. Chem. 55 (7), 3578–3582.

Grael, C.F.F., Albuquerque, S., Lopes, J.L.C., 2005. Chemical constituents of *Lychnophora pohlii* and trypanocidal activity of crude plant extracts and of isolated compounds. Fitoterapia 76, 73–82.

Grael, C.F.F., Vichnewski, W., Petto de Souza, G.E., Callegari Lopes, J.L., Albuquerque, S., Cunha, W.R., 2000. A study of the trypanocidal and analgesic properties from *Lychnophora granmongolense* (Duarte) Semir & Leitao Filho. Phytother. Res. 14, 203–206.

Grecco, S.S., Reimão, J.Q., Tempone, A.G., Sartorelli, P., Romoff, P., Ferreira, M.J.P., Fávero, O.A., Lago, J.H.G., 2010. Isolation of an antileishmanial and antitrypanosomal flavanone from the leaves of *Baccharis retusa* DC. (Asteraceae). Parasitol. Res. 106, 1245–1248.

Guimarães, A.C., Otto, T.D., Alves-Ferreira, M., Miranda, A.B., Degrave, W.M., 2008. In silico reconstruction of the amino acid metabolic pathways of *Trypanosoma cruzi*. Genet. Mol. Res. 7, 872–882.

Hampton, T., 2004. "Promiscuous" anticancer drugs that hit multiple targets may thwart resistance. JAMA 292, 419−422.

Harborne, J.B., 1988. Ecological Biochemistry. Academic Press, London, UK.

Hattori, E.K.O., Nakajima, J.N., 2008. A família Asteraceae na Estação de Pesquisa e Desenvolvimento Ambiental Galheiros, Perdizes, Minas Gerais, Brasil. Rodriguésia 59, 687−749.

Heby, O., Persson, L., Rentala, M., 2007. Targeting the polyamine biosynthetic enzymes: a promising approach to therapy of African sleeping sickness, Chagas' disease and leishmaniasis. Amino Acids 33, 359−366.

Holloway, G.A., Baell, J.B., Fairlamb, A.H., Novello, P.M., Parisot, J.P., Richardson, J., Watson, K.G., Street, I.P., 2007. Discovery of 2- iminobenzimidazoles as a new class of trypanothione reductase inhibitor by high-throughput screening. Bioorg. Med. Chem. Lett. 17 (5), 1422−1427.

Hudock, M.P., Sanz-Rodriguez, C.E., Song, Y., Chan, J.M., Zhang, Y., Odeh, S., Kosztowski, T., Leon-Rossell, A., Concepcion, J.L., Yardley, V., Croft, S.L., Urbina, J.A., Oldfield, E., 2006. Inhibition of *Trypanosoma cruzi* hexokinase by bisphosphonates. J. Med. Chem. 49, 215−223.

Igoillo-Esteve, M., Maugeri, D., Stern, A.L., Beluardi, P., Cazzulo, J.J., 2007. The pentose phosphate pathway in *Trypanosoma cruzi*: a potential target for the chemotherapy of Chagas disease. An. Acad. Bras. Cienc. 79, 649−663.

Iniesta, V., Gomez-Nieto, L.C., Corraliza, I., 2001. The inhibition of arginase by N-omega-hydroxy-l-arginine controls the growth of *Leishmania* inside macrophages. J. Exp. Med. 193 (6), 777−783.

Jansen, R.K., Palmer, J.D., 1987. A DNA chloroplast inversion marks an ancient evolutionary split in the sunflower family (Asteraceae). Proc. Natl. Acad. Sci. U.S.A. 84, 5818−5822.

Jiménez-Jiménez, J., Ledesma, A., Zaragoza, P., González-Barroso, M.M., Rial, E., 2006. Fatty acid activation of the uncoupling proteins requires the presence of the central matrix loop from UCP1. Biochim. Biophys. Acta 1757, 1292−1296.

Jones, D.C., Ariza, A., Chow, W.H.A., Oza, S.L., Fairlamb, A.H., 2010. Comparative structural, kinetic and inhibitor studies of T*rypanosoma brucei* trypanothione reductase with *T-cruzi*. Mol. Biochem. Parasitol. 169 (1), 12−19.

Jordão, C.O., Vichnewski, W., Petto de Souza, G.E., Albuquerque, S., Callegari Lopes, J.L., 2003. Trypanocidal activity of chemical constituents from *Lychnophora salicifolia* Mart. Phytother. Res. 18, 332−334.

Juan, S.M., Cazzulo, J.J., Segura, E.C., 1976. The pyruvate kinase of *Trypanosoma cruzi*. Acta physiol. lat. Am. 26, 424−426.

Jurado, L.A., Machín, I., Urbina, J.A., 1996. *Trypanosoma cruzi* phosphoenolpyruvate carboxykinase (ATP-dependent): transition metal ion requirement for activity and sulfhydryl group reactivity. Biochim. Biophys. Acta 1292, 188−196.

Karioti, A., Skaltsaa, H., Kaiserb, M., Tasdemirc, D., 2009. Trypanocidal, leishmanicidal and cytotoxic effects of anthecotulide-type linear sesquiterpene lactones from *Anthemis auriculata*. Phytomedicine 16, 783−787.

Kirmizibekmez, H., Atay, I., Kaiser, M., Brun, R., Cartagena, M.M., Carballeira, N.M., Yesilada, E., Tasdemir, D., 2011. Antiprotozoal activity of *Melampyrum arvense* and its metabolites. Phytother. Res. 25, 142−146.

Kitamura, E., Otomatsu, T., Maeda, C., Aoki, Y., Ota, C., Misawa, N., Shindo, K., 2013. Production of hydroxlated flavonoids with cytochrome P450 BM3 variant F87V and their antioxidative activities. Biosci. Biotechnol. Biochem. J. 77 (6), 1340−1343.

Krauth-Siegel, R.L., Inhoff, O., 2003. Parasite-specific trypanothione reductase as a drug target molecule. Parasitol. Res. 90, S77−S85.

Krieger, S., Schwarz, W., Ariyanayagam, M.R., Fairlamb, A.H., Krauth-Siegel, R.L., Clayton, C., 2000. Trypanosomes lacking trypanothione reductase are avirulent and show increased sensitivity to oxidative stress. Mol. Microbiol. 35 (3), 542–552.

Lai, D.H., Poropat, E., Pravia, C., Landoni, M., Couto, A.S., Rojo, F.G.P., Fuchs, A.G., Dubin, M., Elingold, I., Rodriguez, J.B., Ferella, M., Esteva, M.I., Bontempi, E.J., Lukes, J., 2014. Solanesyl diphosphate synthase, an enzyme of the ubiquinone synthetic pathway, is required throughout the life cycle of *Trypanosoma brucei*. Eukaryot. Cell 13 (2), 320–328.

Lehane, A.M., Saliba, K.J., 2008. Common dietary flavonoids inhibit the growth of the intraerythrocytic malaria parasite. BMC Res. Notes 1, 26.

Leite, A.C., Neto, A.P., Ambrozin, A.R.P., Fernandes, J.B., Vieira, P.C., Silva, M.F.G.F., Albuquerque, S., 2010. Trypanocidal activity of flavonoids and limonoids isolated from *Myrsinaceae* and *Meliaceae* active plant extracts. Rev. Bras.Farmacogn. 20 (1), 1–6.

Lindoso, J.A.L., Lindoso, A.A.B.P., 2009. Neglected tropical diseases in Brazil. Rev. Inst. Med. Trop. São Paulo 51, 247–253.

Lopes, R.M., Oliveira, T.D., Nagem, T.J., Pinto, A.D.S., 2010. Flavonóides. Biotecnol. Ciênc. Desenvolv. 3 (14), 18–22.

Lu, J., Vodnala, S.K., Gustavsson, A.L., Gustafsson, T.N., Sjoberg, B., Johansson, H.A., Kumar, S., Tjernberg, A., Engman, L., Rottenberg, M.E., Holmgren, A., 2013. Ebsulfur is a benzisothiazolone cytocidal inhibitor targeting the trypanothione reductase of *Trypanosoma brucei*. J. Biol. Chem. 288 (38), 27456–27468.

Maas, M., Hensel, A., Da Costa, F.B., Brun, R., Kaiser, M., Schmidt, T.J., 2011. An unusual dimeric guaianolide with antiprotozoal activity and further sesquiterpene lactones from *Eupatorium perfoliatum*. Phytochemistry 72, 635–644.

Magdesian, M.H., Giordano, R., Ulrich, H., Juliano, M.A., Juliano, L., Schumacher, R.I., Colli, W., Alves, M.J., 2001. Infection by *Trypanosoma cruzi*. Identification of a parasite ligand and its host cell receptor. J. Biol. Chem. 276 (22), 19382–19389.

Magaraci, F., Jimenez, C.J., Rodrigues, C., Rodrigues, J.C., Braga, M.V., Yardley, V., de Luca-Fradley, K., Croft, S.L., de Souza, W., Ruiz-Perez, L.M., Urbina, J., Gonzalez Pacanowska, D., Gilbert, I.H., 2003. Azasterols as inhibitors of sterol 24-methyltransferase in *Leishmania* species and *Trypanosoma cruzi*. J. Med. Chem. 46, 4714–4727.

Manjolin, L.C., Reis, M.B.G., Maquiaveli, C.C., Filho, O.A.S., Silva, E.R., 2013. Dietary flavonoids fisetin, luteolin and their derived compounds inhibit arginase, a central enzyme in *Leishmania (Leishmania) amazonensis* infection. Food Chem. 141 (3), 2253–2262.

Mao, J.H., Mukherjee, S., Zhang, Y., Cao, R., Sanders, J.M., Song, Y.C., Zhang, Y.H., Meints, G.A., Gao, Y.G., Mukkamala, D., Hudock, M.P., Oldfield, E., 2006. Solid-state NMR, crystallographic, and computational investigation of bisphosphonates and farnesyl diphosphate synthase-bisphosphonate complexes. J. Am. Chem. Soc. 128 (45), 14485–14497.

Marín, C., Boutaleb-Charki, S., Díaz, J.G., Huertas, O., Rosales, M.J., Pérez-Cordon, G., Guitierrez-Sánchez, R., Sánchez-Moreno, M., 2009. Antileishmaniasis activity of flavonoids from *Consolida oliveriana*. J. Nat. Prod. 72 (6), 1069–1074.

Martins, G.A.S., Lima, M.D., 2013. Leishmaniose: Do Diagnostico Ao Tratamento. Enciclopédia Biosfera, Centro Científico Conhecer – Goiânia 9 (16), 2566.

Martyn, D.C., Jones, D.C., Fairlamb, A.H., Clardy, J., 2007. High-throughput screening affords novel and selective trypanothione reductase inhibitors with anti-trypanosomal activity. Bioorg. Med. Chem. Lett. 17, 1280–1283.

Maugeri, D.A., Cazzulo, J.J., 2004. The pentose phosphate pathway in *Trypanosoma cruzi*. FEMS Microbiol. Lett. 234, 117–123.

Medda, S., Mukhopadhyay, S., Basu, M.K., 1999. Evaluation of the in-vivo activity and toxicity of amarogentin, an antileishmanial agent, in both liposomal and niosomal forms. J. Antimicrob. Chemother. 44, 791.

Mello, T.T.F.P., Bitencourt, H.R., Pedroso, R.B., Arisstides, S.M.A., Lonardoni, M.V.C., Silveira, T.G.V., 2014. Leishmanicidal activity of synthetic chalcones in *Leishmania (Viannia) braziliensis*. Exp. Parasitol. 136, 27−34.

Miller III, B.R., Roitberg, A.E., 2013. *Trypanosoma cruzi* trans-sialidase as a drug target against Chagas' disease (American trypanosomiasis). Future Med. Chem. 5 (15), 1889−1900.

Ministério da Saúde, 2010a. Guia prático de tratamento da malária no Brasil. Secretaria de Vigilância em Saúde, first ed. Departamento de Vigilância Epidemiológica. Brasília/DF.

Ministério da Saúde, 2010b. Departamento de Ciência e Tecnologia, Secretaria de Ciência, Tecnologia e Insumos Estratégicos. Doenças negligenciadas: estratégias do Ministério da Saúde. Rev. Saúde Pública 44 (1), 200−202.

Ministério da Saúde, 2014. Vigilância Da Esquistossomose Mansoni, fourth ed. Brasília/DF.

Ministério da Saúde, 2016. Manual de Vigilância e Controle da Leishmaniose Visceral. Secretaria de Vigilância em Saúde, first ed. Departamento de Vigilância Epidemiológica. Brasília/DF.

Mishra, B.B., Kale, R.R., Singh, R.K., Tiwari, V.K., 2009. Alkaloids: future prospective to combat leishmaniasis. Fitoterapia 80, 81−90.

Mishra, B.B., Tiwari, V.K., 2011. Natural products: an evolving role in future drug discovery. Eur. J. Med. Chem. 46, 4769−4807.

Muzitano, M.F., Falcão, C.A.B., Cruz, E.A., Bergonzi, M.C., Bilia, A.R., Vincieri, F.F., Rossi-Bergmann, B., Costa, S.S., 2009. Oral metabolism and efficacy of *Kalanchoe pinnata* flavonoids in a murine model of cutaneous leishmaniasis. Planta Méd. 75, 307−311.

Muzitano, M.F., Cruz, E.A., de Almeida, A.P., Da Silva, S.A.G., Kaiser, C.R., Guette, C., Rossi-Bergmann, B., Costa, S.S., 2006a. Quercitrin: Na antileishmanial flavonoid glycoside from *Kalanchoe pinnata*. Planta Méd. 72, 81−83.

Muzitano, M.F., Tinoco, L.W., Guette, C., Kaiser, C.R., Rossi-Bergmann, B., Costa, S.S., 2006b. The antileishmanial activity assessment of unusual flavonoids from *Kalanchoe pinnata*. Phytochemistry 67, 2071−2077.

Montanari, C.A., Bolzani, V.S., 2001. Planejamento racional de fármacos baseado em produtos naturais. Quím. Nova 24 (1), 105−111.

Moore, J.D., Potter, A., 2013. Pin1 inhibitors: pitfalls, progress and cellular pharmacology. Bioorg. Med. Chem. Lett. 23 (15), 4283−4291.

Namdaung, U., Aroonrerk, N., Suksamrarn, S., Danwisetkanjana, K., Saenboonrueng, J., Arjchomphu, W., Suksamrarn, A., 2006. Bioactive constituents of the root bark of *Artocarpus rigidus* subsp. Rigidus. Chem. Pharm. Bull. 54 (10), 1433−1436.

Newman, D.J., Cragg, G.M., 2016. Natural products as sources of new drugs from 1981 to 2014. J. Nat. Prod. 79 (3), 629−661.

Nour, A.M.M., Khalid, S.A., Kaiser, M., Brun, R., Abdalla, W.E., Schmidt, T.J., 2010. The antiprotozoal activity of methylated flavonoids from *Ageratum conyzoides* L. J. Ethnopharmacol. 129, 127−130.

Oduor, R.O., Ojo, K.K., Williams, G.P., Bertelli, F., Mills, J., Maes, L., Pryde, D.C., Parkinson, T., Van Voorhis, W.C., Holler, T.P., 2011. *Trypanosoma brucei* glycogen synthase kinase-3, a target for anti-trypanosomal drug development: a public-private partnership to identify novel leads. PLoS Negl. Trop. Dis. 5 (4), 8.

Ogungbe, I.V., Erwin, W.R., Setzer, W.N., 2014. Antileishmanial phytochemical phenolics: molecular docking to potential protein targets. J. Mol. Graph. Model. 48, 105−117.

Olivares-Illana, V., Pérez-Montfort, R., López-Calahorra, F., Costas, M., Rodríguez-Romero, A., Tuena de Gómez-Puyou, M., Gómez Puyou, A., 2006. Structural differences in triosephosphate isomerase from different species and discovery of a multitrypanosomatid inhibitor. Biochemistry 45, 2556–2560.

Olivares-Illana, V., Rodríguez-Romero, A., Becker, I., Berzunza, M., García, J., Pérez-Montfort, R., Cabrera, N., López-Calahorra, F., Gómez-Puyou, M.T., Gómez-Puyou, A., 2007. Perturbation of the dimer interface of triosephosphate isomerase and its effect on *Trypanosoma cruzi*. PLoS Negl. Trop. Dis. 1, e1.

Ouaissi, M.A., Cornette, J., Capron, A., 1986. Identification and isolation of *Trypanosoma cruzi* trypomastigote cell surface protein with properties expected of a fibronectin receptor. Mol. Biochem. Parasitol. 19, 201–221.

Patterson, S., Alphey, M.S., Jones, D.C., Shanks, E.J., Street, I.P., Frearson, J.A., Wyatt, P.G., Gilbert, I.H., Fairlamb, A.H., 2011. Dihydroquinazolines as a novel class of *Trypanosoma brucei* trypanothione reductase inhibitors: discovery, synthesis, and characterization of their binding mode by protein crystallography. J. Med. Chem. 54 (19), 6514–6530.

Pratt, C., Nguyen, S., Phillips, M.A., 2014. Genetic validation of *Trypanosoma brucei* glutathione synthetase as an essential enzyme. Eukaryot. Cell 13 (5), 614–624.

Pérez-Montfort, R., Garza-Ramos, G., Alcántara, G.H., Reyes-Vivas, H., Gao, X.G., Maldonado, E., de Gómez-Puyou, M.T., Gómez-Puyou, A., 1999. Derivatization of the interface cysteine of triosephosphate isomerase from *Trypanosoma brucei* and *Trypanosoma cruzi* as probe of the interrelationship between the catalytic sites and the dimer interface. Biochemistry 38, 4114–4120.

Phillips, C., Dohnalek, J., Gover, S., Barrett, M.P., Adams, M.J., 1998. A 2.8 angstrom resolution structure of 6-phosphogluconate dehydrogenase from the protozoan parasite *Trypanosoma brucei*: comparison with the sheep enzyme accounts for differences in activity with coenzyme and substrate analogues. J. Mol. Biol. 282 (3), 667–681.

Pierce, R.J., Dubois-Abdesselem, F., Caby, S., Trolet, J., Lancelot, J., Oger, F., Bertheaume, N., Roger, E., 2011. Chromatin regulation in schistosomes and histone modifying enzymes as drug targets. Mem. Inst. Oswaldo Cruz 106 (7), 794–801.

Pordeus, L.C., Aguiar, L.R., Quinino, L.R.M., Barbosa, C.S., 2008. The occurrence of acute and chronic forms of the Schistossomiais Mansonic in Brazil from 1997 to 2006: a revision of literature. Epidemiol. Serv. Saúde 17 (3), 163–175.

Pourshafie, M., Morand, S., Virion, A., Rakotomanga, M., Dupuy, C., Loiseau, P.M., 2004. Cloning of S-adenosyl-L-methionine: C-24-Delta-sterolmethyltransferase (ERG6) from Leishmania donovani and characterization of mRNAs in wild-type and amphotericin B-Resistant promastigotes. Antimicrob. Agents Chemother. 48, 2409–2414.

Racagni, G.E., Machado de Domenech, E.E., 1983. Characterization of *Trypanosoma cruzi* hexokinase. Mol. Biochem. Parasitol. 9, 181–188.

Rackham, M.D., Brannigan, J.A., Rangachari, K., Meister, S., Wilkinson, A.J., Holder, A.A., Leatherbarrow, R.J., Tate, E.W., 2014. Design and synthesis of high affinity inhibitors of *Plasmodium falciparum* and *Plasmodium vivax* *N*-myristoyltransferases directed by ligand efficiency dependent lipophilicity (LELP). J. Med. Chem. 57 (6), 2773–2788.

Reyes-Vivas, H., Hernández-Alcantara, G., López-Velazquez, G., Cabrera, N., Pérez-Montfort, R., de Gómez-Puyou, M.T., Gómez-Puyou, A., 2001. Factors that control the reactivity of the interface cysteine of triosephosphate isomerase from *Trypanosoma brucei* and *Trypanosoma cruzi*. Biochemistry 40, 3134–3140.

Ribeiro, J.N., Oliveira, T.T.D., Nagem, T.J., Ferreira Júnior, D.B., Pinto, A.D.S., 2006. Avaliação dos parâmetros sangüíneos de hepatotoxicicidade em coelhos normais submetidos a tratamentos com antocianina e antocianina + naringenina. Rev. Bras. Anál. Clín. 38 (1), 23–27.

Ribeiro, A., Piló-Veloso, D., Romanha, A.J., Zani, C.L., 1997. Trypanocidal flavonoids from *Trixis vauthieri*. J. Nat. Prod. 60, 836–838.

Roberts, C.W., McLeod, R., Rice, D.W., Ginger, M., Chance, M.L., Goad, L.J., 2003. Fatty acid and sterol metabolism: potential antimicrobial targets in apicomplexan and trypanosomatid parasitic protozoa. Mol. Biochem. Parasitol. 126, 129–142.

Rodrigues Da Silva, L., et al., 2015. Flavonóides: constituição química, ações medicinais e potencial tóxico. Acta Toxicol. Argent. Ciudad Autónoma de Buenos Aires, 23 (1), 36–43.

Rodríguez-Romero, A., Hernández-Santoyo, A., del Pozo Yauner, L., Kornhauser, A., Fernández-Velasco, D.A., 2002. Structure and inactivation of triosephosphate isomerase from *Entamoeba histolytica*. J Mol. Biol. 322, 669–675.

Roque, N., Bautista, H., 2008. Asteraceae – Caracterização e Morfologia Floral. Universidade Federal da Bahia, Salvador.

Salmanpour, R., Razmavar, M.R., Abtahi, N., 2006. Comparison of intralesional meglumine antimoniate, cryotherapy and their combination in the treatment of cutaneous leishmaniasis. Int. J. Dermatol. 45, 1115–1116. http://dx.doi.org/10.1111/j.1365-4632.2006.02822.x.

Sanz-Rodriguez, C.E., Concepcion, J.L., Pekerar, S., Oldfield, E., Urbina, J.A., 2007. Bisphosphonates as inhibitors of *Trypanosoma cruzi* hexokinase: kinetic and metabolic studies. J. Biol. Chem. 282, 12377–12387.

Schmidt, T.J., Khalid, A.S., Romanha, J.A., Alves, T.M.A., Biavatti, M.W., Brun, R., Da Costa, B.F., De Castro, S.L., Ferreira, V.F., De Lacerda, V.G.M., Lago, J.H.G., Leon, L.L., Lopes, N.P., Das Neves, A.C.R., Niehues, M., Ogungbe, I.V., Pohlit, A.M., Scotti, M.T., Setzer, W.N., Soeiro, M.N.C., Steindel, M., Tempone, A.G., 2012. The potential of secondary metabolites from plants as drugs or leads against protozoan neglected diseases – Part II. Curr. Med. Chem. 19, 2176–2228.

Scotti, L., Ishiki, H., Mendonca, F.J.B., Da Silva, M.S., Scotti, M.T., 2015. In-silico analyses of natural products on leishmania enzyme targets. Mini Rev. Med. Chem. 15 (3), 253–269. ISSN:1389-5575/1875-5607. http://dx.doi.org/10.2174/1389557515031503121141854.

Scotti, L., Mendonça, F.J., da Silva, M.S., Scotti, M.T., 2016. Enzymatic targets in *Trypanosoma brucei*. Curr. Protein Pept. Sci. 17 (3), 243–259.

Sen, R., Bandyopadhyay, S., Dutta, A., Mandal, G., Ganguly, S., Saha, P., Chatterjee, M., 2007. Artemisinin triggers induction of cell-cycle arrest and apoptosis in *Leishmania donovani* promastigotes. J. Med. Microbiol. 56, 1213.

Setzer, W.N., 2013. Trypanosomatid disease drug discovery and target identification. Future Med. Chem. 5 (15), 1703–1704.

Shaw, M.P., Bond, C.S., Roper, J.R., Gourley, D.G., Ferguson, M.A.J., Hunter, W.N., 2003. High-resolution crystal structure of *Trypanosoma brucei* UDP-galactose 4′-epimerase: a potential target for structure-based development of novel trypanocides. Mol. Biochem. Parasitol. 126 (2), 173–180.

Silber, A.M., Colli, W., Ulrich, H., Alves, M.J., Pereira, C.A., 2005. Amino acid metabolic routes in *Trypanosoma cruzi*: possible therapeutic targets against Chagas' disease. Curr. Drug Targets Infect. Disord. 5, 53–64.

Singh, N., Misra, K., 2009. Computational screening of molecular targets in *Plasmodium* for novel non resistant anti-malarial drugs. Bioinformation 3 (6), 255–262.

Soares-Bezerra, R.J., Leon, L., Genstra, M., 2004. Recentes avanços da quimioterapia das leishmanioses: moléculas intracelulares como alvo de fármacos. Braz. J. Pharm. Sci. 40, 139–149.

Sousa, F.S., Ruiz, E.E.S., 2015. Aplicação da teoria de redes complexas no estudo de relacionamento entre doenças em casos de óbito do paciente. Universidade de São Paulo (USP), Brasil.

Steverding, D., 2010. The development of drugs for treatment of sleeping sickness: a historical review. Parasit. Vectors 3, 9.

Stuessy, T.F., Bohm, B.A., 2001. Flavonoids of the Sunflower Family (Asteraceae). Springer, Vienna.

Sülsen, V.P., Cazorla, S.I., Frank, F.M., Redko, F.C., Anesini, C.A., Coussio, J.D., Malchiodi, E.L., Martino, V.S., Muschietti, L.V., 2007. Trypanocidal and leishmanicidal activities of flavonoids from Argentine medicinal plants. Am. J. Trop. Med. Hyg. 77, 654–659.

Sylvester, D., Krassner, S.M., 1976. Proline metabolism in *Trypanosoma cruzi* epimastigotes. Comp. Biochem. Physiol. B 55, 443–447.

Tang, W., Eisenbrand, G., 1992. Springer-Verlag, Berlin, p. 160.

Taleb-Contini, S.H., Salvador, M.J., Balanco, J.M.F., Albuquerque, S., de Oliveira, D.C.R., 2004. Antiprotozoal effect of crude extracts and flavonoids isolated from *Chromolaena hirsuta* (Asteraceae). Phytother. Res. 18, 250–254.

Tasdemir, D., Kaiser, M., Brun, R., Yardley, R., Schmidt, T.J., Tosun, F., Rüedi, P., 2006. Antitrypanosomal and antileishmanial activities of flavonoids and their analogues: in vitro, in vivo, structure-activity relationship, and quantitative structure-activity relationship studies. Antimicrob. Agents Chemother. 50, 1352–1364.

Tate, E.W., Bell, A.S., Rackham, M.D., Wright, M.H., 2014. *N*-myristoyltransferase as a potential drug target in malaria and leishmaniasis. Parasitology 141 (1), 37–49.

Tetaud, E., Bringaud, F., Chabas, S., Barrett, M.P., Baltz, T., 1994. Characterization of glucose transport and cloning of a hexose transporter gene in *Trypanosoma cruzi*. Proc. Natl. Acad. Sci. U.S.A 91, 8278–8282.

Tielens, A.G., Van der Meer, P., van den Heuvel, J.M., van den Bergh, S.G., 1991. The enigmatic presence of all gluconeogenic enzymes in *Schistosoma mansoni* adults. Parasitology 102 (2), 267–276.

Trapani, S., Linss, J., Goldenberg, S., Fischer, H., Craievich, A.F., Oliva, G., 2001. Crystal structure of the dimeric phosphoenolpyruvate carboxykinase (PEPCK) from *Trypanosoma cruzi* at 2 A resolution. J. Mol. Biol. 313, 1059–1072.

Urbina, J.A., Machin, I., Jurado, L., 1993. The limitations of paradigms: studies on the intermediary metabolism of *Trypanosoma cruzi*. Biol. Res. 26, 81–88.

Urbina, J.A., Osorno, C.E., Rojas, A., 1990. Inhibition of phosphoenolpyruvate carboxykinase from *Trypanosoma (Schizotrypanum) cruzi* epimastigotes by 3-mercaptopicolinic acid: in vitro and in vivo studies. Arch. Biochem. Biophys. 282, 91–99.

Urbina, J.A., Crespo, A., 1984. Regulation of energy metabolism in *Trypanosoma (Schizotrypanum) cruzi* epimastigotes. I. Hexokinase and phosphofructokinase. Mol. Biochem. Parasitol. 11, 225–239.

van Hellemond, J.J., Tielens, A.G., 2006. Adaptations in the lipid metabolism of the protozoan parasite *Trypanosoma brucei*. FEBS Lett. 580, 5552–5558.

Venugopal, V., Datta, A.K., Bhattacharyya, D., Dasgupta, D., Banerjee, R., 2009. Structure of cyclophilin from *Leishmania donovani* bound to cyclosporin at 2.6Å resolution: correlation between structure and thermodynamic data. Acta Crystallogr. D65, 1187–1195.

Vitalini, S., Beretta, G., Iriti, M., Orsenigo, S., Basilico, N., Dall'Acqua, S., Iorizzi, M., Fico, G., 2011. Phenolic compounds from *Achillea millefolium* L. and their bioactivity. Acta Biochim. Pol. 58, 203–209.

Wheatley, N.C., Andrews, K.T., Tran, T.L., Lucke, A.J., Reid, R.C., Fairlie, D.P., 2010. Antimalarial histone deacetylase inhibitors containing cinnamate or NSAID components. Bioorg. Med. Chem. Lett. 20, 7080–7084.

Williams, C., Espinosa, O.A., Montenegro, H., Cubilla, L., Capson, T.L., Ortega-Barn, E., Romero, L.I., 2003. Hydrosoluble formazan XTT: its application to natural products drug discovery for *Leishmania*. J. Microbiol. Methods 55 (3), 813–816.

Wiemer, E.A., IJlst, L., van Roy, J., Wanders, R.J., Opperdoes, F.R., 1996. Identification of 2-enoyl coenzyme A hydratase and NADP(+)- dependent 3-hydroxyacyl-CoA dehydrogenase activity in glycosomes of procyclic *Trypanosoma brucei*. Mol. Biochem. Parasitol. 82, 107–111.

Wright, M.H., Clough, B., Rackham, M.D., Rangachari, K., Brannigan, J.A., Grainger, M., Moss, D.K., Bottrill, A.R., Heal, W.P., Broncel, M., Serwa, R.A., Brady, D., Mann, D.J., Leatherbarrow, R.J., Tewari, R., Wilkinson, A.J., Holder, A.A., Tate, E.W., 2014. Validation of *N*-myristoyltransferase as an antimalarial drug target using an integrated chemical biology approach. Nat. Chem. 6 (2), 112–121.

Zelck, U.E., Von Janowsky, B., 2004. Antioxidant enzymes in intramolluscan *Schistosoma mansoni* and ROS-induced changes in expression. Parasitology 128 (Pt5), 493–501.

Zimmermann, S., Oufir, M., Leroux, A., Krauth-Siegel, R.L., Becker, K., Kaiser, M., Brun, R., Hamburger, M., Adams, M., 2013. Cynaropicrin targets the trypanothione redox system in *Trypanosoma brucei*. Bioorg. Med. Chem. 21 (22), 7202–7209.

Chapter 8

Quasi-SMILES as a Novel Tool for Prediction of Nanomaterials' Endpoints

A.P. Toropova[1], A.A. Toropov[1], A.M. Veselinović[2], J.B. Veselinović[2], D. Leszczynska[3], J. Leszczynski[3]
[1]*IRCCS-Istituto di Ricerche Farmacologiche Mario Negri, Milano, Italy;* [2]*University of Niš, Niš, Serbia;* [3]*Jackson State University, Jackson, MS, United States*

8.1. INTRODUCTION

The influence of various nanomaterials on everyday life gradually increases owing to their potential of being useful for different applications in medicine (De Jong and Borm, 2008; Webster et al., 2013; Toropova et al., 2016a,b).

As a rule, generally, experimental measurement of an endpoint is not cheap. In addition, performing the experiment demands considerable time. This promotes the development of alternative techniques, able to provide investigated data faster and more efficiently. Such techniques are available in a large pool of computational chemistry methods. Specifically, the techniques of calculations of endpoints, which are able to include experimental data related to untested but similar substances, become attractive alternatives for the experiment. Quantitative structure—property/activity relationships (QSPRs/QSARs) represent the practical application of the aforementioned alternatives. The theory and praxis of QSPR/QSAR have an impressive record of successful utilizations for the prediction of endpoints related to organic (Toropova et al., 2011a), inorganic (Toropova et al., 2011b), organometallic (Toropova et al., 2011c), and polymeric (Duchowicz et al., 2015) species.

The evolution of the QSPR/QSAR theory/praxis involves a few components. It benefits the improvement of algorithms for the analysis of available data, which allows the prediction of the physicochemical and/or biochemical behavior of substances that were not examined in the experiment. The second, less discussed component is of equal importance. It involves the establishment of the definition of the task (target). Historically, the development of correlations "descriptor—endpoint" for a sole endpoint was the main aim of

Multi-Scale Approaches in Drug Discovery. http://dx.doi.org/10.1016/B978-0-08-101129-4.00008-4

QSPR/QSAR modeling in the beginning of the applications of this approach (Wiener, 1947a,b, 1948; Gutman et al., 2005, 2009; Hosoya, 1972; Bonchev et al., 1980). Later, the QSPR/QSAR analysis aimed at more challenging tasks—prediction of not a single property but a group of important and sometimes interdependent endpoints (Speck-Planche et al., 2011, 2012a,b, 2013).

The classes of substances considered for the QSPR/QSAR analysis have broadened over the years. An important impetus for the development of novel approaches arrived after the applications of the classical QSAR methodology to nanomaterials—unique class of chemical species—failed. The growing importance of these species is illustrated by the fast growing number of publications dedicated to nanomaterials. In 2000, there was about 100 articles related to the keyword "nanomaterial." This number has expanded to 11,000 in 2015 (Fig. 8.1).

Obviously, there have been numerous attempts to utilize the QSPR/QSAR approach for nanomaterials with the application of various "nanodescriptors" (Oksel et al., 2015). However, approaches focused on building up "nano-QSAR" were based on hardly accessible physicochemical characteristics of nanomaterials (Sayes and Ivanov, 2010; Glotzer and Solomon, 2007). Interestingly, the traditional descriptors appropriate for substances that are not nanomaterials were also examined as a tool to build up "nano-QSAR" (Fourches et al., 2011). However, in this work, all nanoparticles had the same "nano" metal core, and the difference between nanoparticles was defined solely by small organic molecules (Fourches et al., 2011). Naturally, for such species, traditional descriptors can be quite appropriate.

The so-called optimal descriptors provide the possibility to build a predictive model for various nanomaterials using *all* available eclectic data represented by quasi—simplified molecular input-line entry system (SMILES). This holistic approach was successfully applied for the development of model for membrane damage by ZnO and TiO_2 nanoparticles (Toropova and Toropov, 2013; Toropova et al., 2014), mutagenicity of fullerene (Toropov and

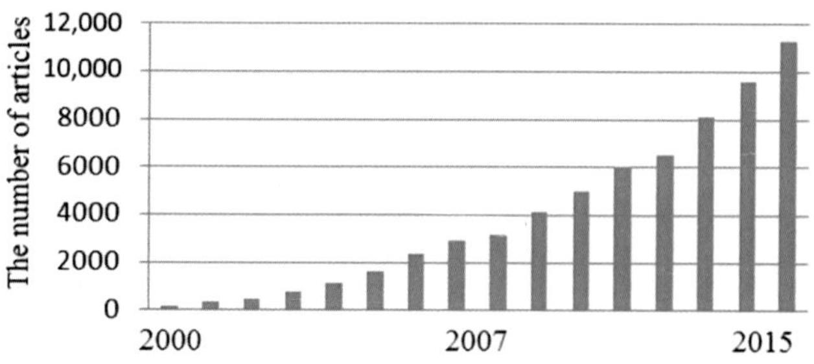

FIGURE 8.1 The increase of numbers of articles related to keyword "nanomaterial" (2000–2015), according to www.sciencedirect.com.

Toropova, 2014), mutagenic potential of multiwalled carbon nanotubes (Toropov and Toropova, 2015), and cytotoxicity of metal oxide nanoparticles to bacteria *Escherichia coli* (Toropova et al., 2012).

We believe that the increasing interest in the application of efficient methods that reliably predict the characteristics of nanomaterials justifies the review of such approaches and the results obtained using those techniques. The aims of this chapter are (1) description of the method of building up quasi-SMILES and (2) introduction of the principles of development of predictive models based on quasi-SMILES. The second aim could be efficiently accomplished using the CORAL software available on the Internet (CORAL, http://www.insilico.eu/coral). The readers are welcome to carry out their own research projects using the CORAL program.

It should be noted that nowadays, the quasi-SMILES constitute a promising alternative in a situation in which one should construct a model based on eclectic data, such as physicochemical and biochemical conditions of a phenomena, presence of large number of factors that can impact the phenomena, and uncertainty in the correctness of classification of all factors into (1) those with significant impact and (2) those with negligible influence. Unfortunately, the aforementioned indeterminacy often takes place in various stages of drug discovery.

8.2. METHOD

8.2.1. SMILES and Quasi-SMILES

SMILES has been introduced by Weiniger and collaborators (Weininger, 1988, 1990; Weininger et al., 1989). This approach allows for simple representation of the molecular structures.

There are defined equivalences between the representation of the molecular structure by graphs and using SMILES approach. However, one also needs to be aware about their significant distinctions. These are reviewed in Toropov et al., 2011.

Optimal descriptors have been improved along with advances of QSAR approaches. During the initial steps of evolution of the optimal descriptors the molecular graph was the basis for building up a QSAR model. A very similar (if not identical) approach has been developed for SMILES and SMILES attributes. It can be summarized as follows:

1. Each SMILES of the training set provides a list of attributes, x_{kj} (Toropov et al., 2011):

$$\text{SMILES}_k \rightarrow \{x_{k1}, x_{k2}, \ldots, x_{km}\} \tag{8.1}$$

2. The Monte Carlo method provides correlation weights for the total list of attributes. They are extracted from all SMILES notations of the training set, which give the maximal correlation coefficient between the examined endpoint and sums of correlation weights for SMILES of the training set:

$$Monte\_Carlo\_method \rightarrow \{CW(x_{k1}), CW(x_{k2}), ..., CW(x_{km})\} \qquad (8.2)$$

3. The predictive model is represented by a one-variable linear equation:

$$\begin{aligned}&Least\_squares\_method \rightarrow \\ &EP_k = C_0 + C_1 \times \sum_{x_{kj} \in \text{SMILES}} CW(x_{kj}) = C_0 + C_1 \times \text{DCW}(T^*, N^*)\end{aligned} \qquad (8.3)$$

In the vector and matrix representations, this approach can be expressed as the following:

$$\begin{pmatrix} MS_1 \\ MS_2 \\ \cdots \\ MS_n \end{pmatrix} \rightarrow \begin{bmatrix} x_{11} & x_{12} & \cdots & x_{1m} \\ x_{21} & x_{22} & \cdots & x_{2m} \\ \cdots & \cdots & \cdots & \cdots \\ x_{n1} & x_{n2} & \cdots & x_{nm} \end{bmatrix} \leftrightarrow \begin{pmatrix} E_1 \\ E_2 \\ \cdots \\ E_n \end{pmatrix} \qquad (8.4)$$

where MS_k are molecular structures (represented by graph or SMILES) and x_{kj} represent molecular features extracted from molecular graphs or molecular features extracted from SMILES. However, an endpoint can be interdependent with respect to some additional impacts related to physicochemical and/or biochemical conditions. In this case, instead of traditional SMILES, one should utilize an extension of the classical parameters referred to as quasi-SMILES. The basis of building up quasi-SMILES can be extracted from a graph, SMILES, and additional eclectic data.

In the traditional approach, one assumes that an endpoint depends on the molecular structure. However, there are cases in which this approach has to be revised. Obviously, there are also situations in which one can expect that the endpoint depends on other conditions (temperature, concentration, dose, etc.) and/or circumstances (the presence/absence of illumination, magnetic field, different times of exposure, etc.). In this case, instead of the paradigm:

"Endpoint = F (Molecular Structure)"
one should apply other paradigm:
"Endpoint = F (Eclectic Data)".

The quasi-SMILES is a representation of the eclectic data. The aforementioned scheme (1)-(2)-(3) is represented by the quasi-SMILES (eclectic data), ED_k, correlation weights of symbols from quasi-SMILES, $CW(x_{kj})$, and experimental data obtained for thc studied endpoint, E_k:

$$\begin{pmatrix} ED_1 \\ ED_2 \\ \ldots \\ ED_n \end{pmatrix} \rightarrow \begin{bmatrix} CW(x_{11}) & CW(x_{12}) & \ldots & CW(x_{1m}) \\ CW(x_{21}) & CW(x_{22}) & \ldots & CW(x_{2m}) \\ \ldots & \ldots & \ldots & \ldots \\ CW(x_{n1}) & CW(x_{n2}) & \ldots & CW(x_{nm}) \end{bmatrix} \leftrightarrow \begin{pmatrix} E_1 \\ E_2 \\ \ldots \\ E_n \end{pmatrix} \tag{8.5}$$

Thus, the vector of eclectic data represents quasi-SMILES. The quasi-SMILES is a string of symbols similar to traditional SMILES, but the meaning of each symbol in quasi-SMILES is not necessarily the representation of molecular features. Fig. 8.2 shows the general scheme of utilization of quasi-SMILES.

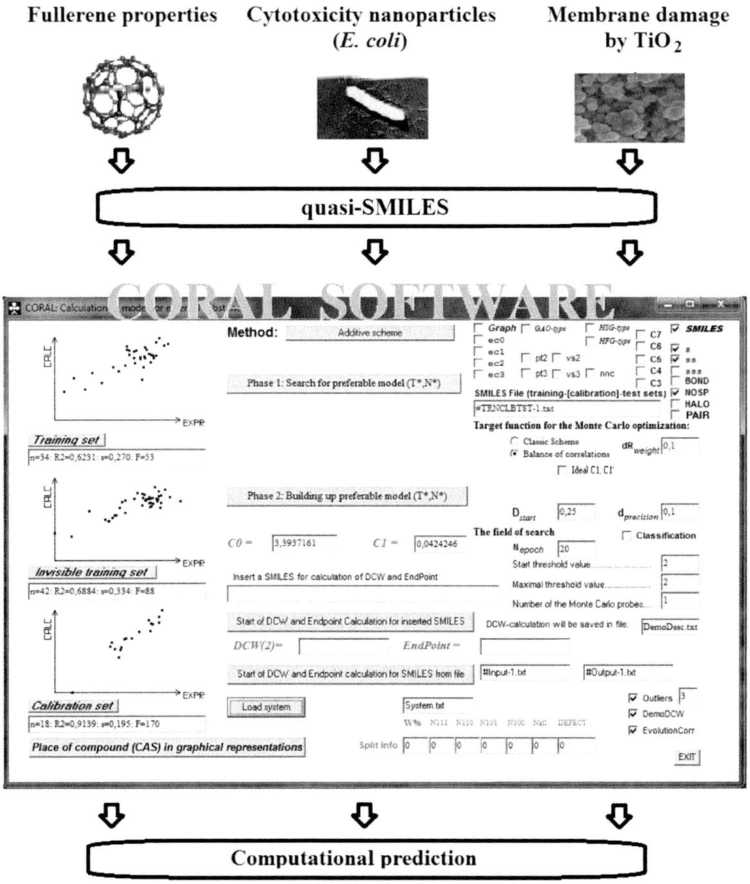

FIGURE 8.2 The general scheme of utilization of quasi–simplified molecular input-line entry system.

8.2.2. Monte Carlo Method

In the case represented by Eq. (8.5), the Monte Carlo method is used to optimize the correlation weights $CW(x_{kj})$. Here, the target function represents the correlation coefficient between the endpoint values E_k and the sum of correlation weights of symbols from the corresponding quasi-SMILES extracted from the training set.

The sequence of modification of correlation weights is random for each epoch of the Monte Carlo optimization. The epoch represents step-by-step modification of all correlation weights involved in building up a QSAR model. It is to be noted that one should define a threshold to classify the symbols of quasi-SMILES into two classes: (1) rare and (2) not rare. The correlation weights of rare symbols are fixed to 0, and consequently, they are not involved in building up a model.

In the case of unlimited number of epochs of the Monte Carlo optimization, the probability of overtraining is very high. Under such circumstances, instead of using an unlimited number, it is better to use the number of epochs that gives a preferable statistics for a calibration set. In principle, the measures of the statistical quality of the calibration set can be: (1) correlation coefficient between experimental and calculated values of an endpoint and (2) root-mean squared error. The graphical illustration (Fig. 8.3) shows that these two approaches can give different values of the preferable number of epochs. The computational experiments indicate that the correlation coefficient provides a more reliable criterion, because this it often gives preferable predictive potential for an external invisible validation set. However, in addition to the preferable number of epochs, one should also select a preferable threshold (T*). Thus, the goal is the selection of a satisfactory pair of values: T = T* and N=N*, which gives preferable statistical quality for the calibration set (Fig. 8.4).

8.2.3. Utilization of the Model

The result of the Monte Carlo optimization provides the list of correlation weights for symbols involved in the model. Each symbol is a representation of defined circumstance. For instance, (1) temperature range, i.e., 100−110°C can be denoted as a code "a," 110−120°C denoted as code "b," etc.; (2) dose ranges, i.e., 20−25 mg/kg is denoted as "c," 25−30 mg/kg is denoted as "d", etc.; (3) time of exposure 1 h is denoted as "e," 2 h is denoted as "f," and so on, according to the corresponding conditions and circumstances.

Having the data on the correlation weights, one can extract a list of symbols from the corresponding quasi-SMILES and calculate:

$$\mathrm{DCW}(T^*, N^*) = \sum_{x_{kj} \in \text{quasi-SMILES}} CW(x_{kj}) \tag{8.6}$$

$$EP_k = C_0 + C_1 \times \mathrm{DCW}(T^*, N^*) \tag{8.7}$$

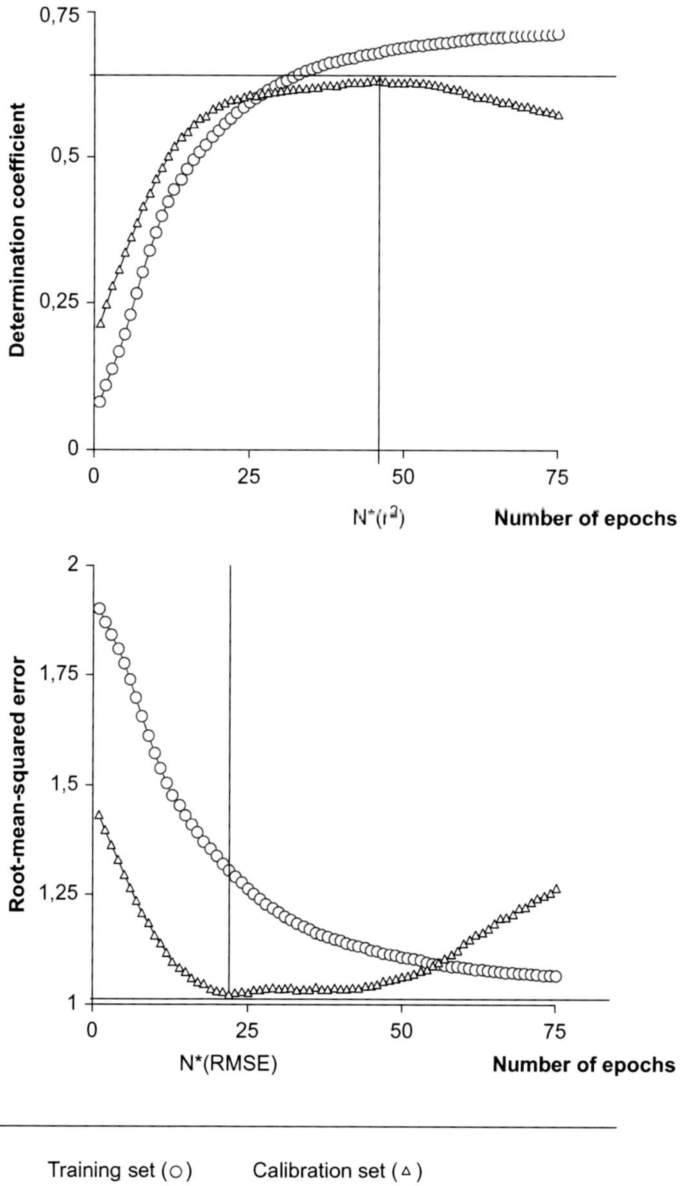

FIGURE 8.3 The selection of the number of epochs of the Monte Carlo optimization using (1) the correlation coefficient between experimental and predicted values of an endpoint and (2) root-mean squared error.

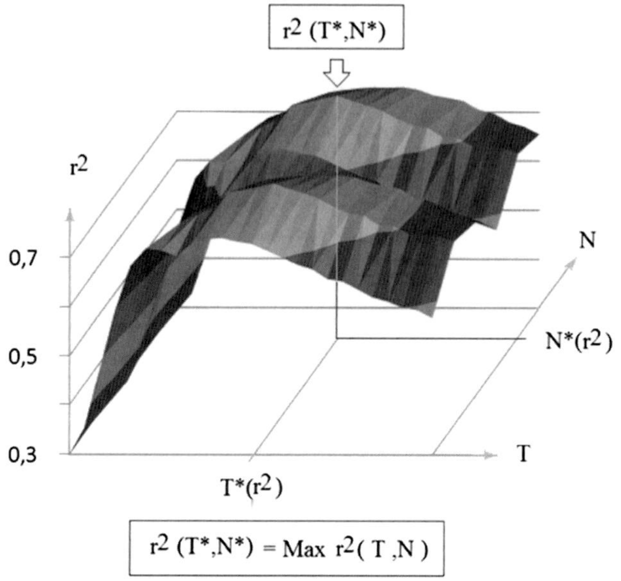

FIGURE 8.4 The scheme of the definition of the T* and N* values that give preferable statistics for the calibration set.

8.2.4. Domain of Applicability

The experimental data are used for model development and for evaluation of the model quality. The splitting of data into the "visible" training set (for the described approach the "visible" training set also contains the calibration set) and "invisible" validation set has apparent influence on the predictability of a model. A possible measure of the quality of the split can be estimated from the prevalence of each feature in the training and calibration sets:

$$defect(x_{kj}) = \sum_{active} |P(x_{kj}) - P\prime(x_{kj})| \tag{8.8}$$

where the probability of the feature x_{kj} in the training set $P(x_{kj})$ and the probability of x_{kj} in the calibration set $P(x_{kj})$ are calculated by:

$$P(x_{kj}) = \frac{N_{set}(x_{kj})}{N_{set}} \tag{8.9}$$

where $N_{set}(x_{kj})$ is the number of quasi-SMILES that contains x_{kj} and N_{set} represents the total number of quasi-SMILES in the set. The quality of split is evaluated based on defect value. The defect is calculated with active (not blocked) x_{kj} only. If the defect $= 0$, the split should be considered as an "ideal" one. However, in fact, this situation is not possible. The value of the defect calculated with Eq. (8.8) gives possibility to compare the quality of various splits.

The sum of *defects* (x_{kj}) of all active attributes of quasi-SMILES can be a measure of a defect of each quasi-SMILES:

$$defect(\text{quasi\_SMILES}_k) = \sum_{x_{kj} \in \text{quasi\_SMILES}_k} defect(x_{kj}) \tag{8.10}$$

Summation of all *defects (*quasi_SMILES*)* can be considered as a measure of the quality of split of data into the visible training, calibration, and invisible validation sets:

$$defect(split) = \sum_{\text{quasi\_SMILES}_k \in Training} defect(\text{quasi\_SMILES}_k) \tag{8.11}$$

The probabilistic domain of applicability can be defined via inequality:

$$defect(\text{quasi\_SMILES}) < 2 \times \overline{defect(\text{quasi\_SMILES})} \tag{8.12}$$

In other words, if quasi-SMILES characterized by the *defect (*quasi-SMILES*)*, which is lower than the doubled average value of this characteristics over compounds, is included in the training set, then this quasi-SMILES falls into the domain of applicability. Otherwise, this quasi-SMILES is outside the domain of applicability. In addition, one can compare two splits using the defect (split) calculated with Eq. (8.11). The split characterized by lower defect is better.

8.2.5. Mechanistic Interpretation

The described approach allows defining the mechanical interpretation of model based on the correlation weights of active features extracted from quasi-SMILES. Having the numerical data on the correlation weights of features that takes place in several runs of the Monte Carlo optimization, one can extract three categories of these features:

1. Features that have positive values of the correlation weight in all runs. These are promoters of endpoint increase.
2. Features that have negative values of the correlation weight in all runs. These are promoters of endpoint decrease.
3. Features that have both negative and positive values of the correlation weight in different runs of the optimization. These are features with unclear role (one cannot classify these features as promoters of increase or decrease for endpoint).

8.3. EXAMPLES OF APPLICATIONS OF QUASI-SMILES FOR NANOMATERIALS

The principles of model development and selection of descriptors discussed in the previous sections have been tested in various cases. Examples of such studies are provided in the next few sections.

8.3.1. Format of Representation of a Model

The format of representation of a predictive QSAR model represents an extremely important feature for a potential user of the model. There are well-known OECD (Organization for Economic Co-operation and Development) principles widely used in the QSPR/QSAR analyses. However, in the case of the model based on the quasi-SMILES, the scheme of building up quasi-SMILES involves additional information. Thus, the format of representation of a model used in this work is the following:

- The description of endpoint;
- The description of quasi-SMILES;
- The statistical characteristics of model;
- Domain of applicability;
- Mechanistic interpretation.

The general scheme of the algorithm for building up a model is described in the section "Method."

8.3.2. Cytotoxicity for Metal Oxide Nanoparticles Under Different Conditions

8.3.2.1. The Description of Endpoint

The numerical data on cytotoxicity of metal oxide nanoparticles to *Escherichia coli* (the concentration of the nanoparticles that proved to be lethal to 50% of the bacteria *E. coli* LC_{50}, in moles per liter) have been taken from the literature (Pathakoti et al., 2014). The negative decimal logarithm of the LC_{50} (pLC_{50}) has been considered as the endpoint. The dark cytotoxicity and photo-induced cytotoxicity were examined as united endpoint, owing to the application of the model that is a mathematical function of the atomic composition and conditions (the presence/absence of photoinducing).

8.3.2.2. The Description of Quasi-SMILES

In the case of cytotoxicity in darkness, traditional SMILES were used to represent metal oxide nanoparticles. In the case of photo-induced cytotoxicity, the symbol "^" was added at the end of the traditional SMILES. Thus, absence of "^" means the acting of nanoparticle in darkness, and the presence of "^" means the acting of nanoparticle under illumination (Table 8.1) (Toropova et al., 2015).

8.3.2.3. The Statistical Characteristics of Model

The best model for cytoctoxicity of metal oxide nanoparticles based on quasi-SMILES (Toropova et al., 2015) is the following:

$$
\begin{aligned}
&pLC_{50} = 1.5185(\pm 0.0334) + 0.8370(\pm 0.0110) \times DCW(1,9)\\
&n = 22, r^2 = 0.9081, s = 0.354, F = 198(\text{training set})\\
&n = 6, r^2 = 0.9943, s = 0.454(\text{calibration set})\\
&n = 6, r^2 = 0.9835, s = 0.418(\text{validation set})
\end{aligned}
\tag{8.13}
$$

TABLE 8.1 Quasi-SMILES Used to Build up Model for Cytotoxicity of Metal Oxide Nanoparticles

Splits[a]						Quasi-SMILES for metal oxide nanoparticles	pLC_{50} in mol/L [30]
1	2	3	4	5	6		
v	t	c	t	t	c	O=[Zn]	5.80
t	c	v	t	t	t	[Cu]=O	4.24
t	t	c	t	c	t	O=[V]O[V]=O	3.48
c	t	t	c	t	t	O=[Y]O[Y]=O	5.79
t	c	c	t	t	t	O=[Bi]O[Bi]=O	3.55
t	t	t	t	v	c	O=[In]O[In]=O	2.83
t	c	t	t	c	t	O=[Sb]O[Sb]=O	3.12
t	v	c	v	v	v	O=[Al]O[Al]=O	2.42
t	t	t	c	t	v	O=[Fe]O[Fe]=O	2.40
c	t	v	t	t	t	O=[Si]=O	2.54
v	c	v	v	t	c	O=[Zr]=O	2.58
t	t	t	v	v	c	O=[Sn]=O	2.53
t	t	t	t	t	t	O=[Ti]=O	2.14
t	t	t	c	t	t	[Co]=O	3.13
t	v	t	t	c	t	[Ni]=O	3.79
v	c	c	t	c	c	O=[Cr]O[Cr]=O	2.06
t	t	t	v	t	v	O=[La]O[La]=O	4.96
t	t	t	c	c	t	O=[Zn]^	6.23
t	t	t	t	t	t	[Cu]=O^	5.71
c	c	t	t	t	t	O=[V]O[V]=O^	3.78
t	v	c	t	v	t	O=[Y]O[Y]=O^	5.84
c	t	t	t	t	c	O=[Bi]O[Bi]=O^	4.02
t	t	v	v	t	c	O=[In]O[In]=O^	3.48
v	t	t	c	t	t	O=[Sb]O[Sb]=O^	3.66
t	v	t	v	t	t	O=[Al]O[Al]=O^	2.75
t	t	v	t	t	v	O=[Fe]O[Fe]=O^	2.54

Continued

TABLE 8.1 Quasi-SMILES Used to Build up Model for Cytotoxicity of Metal Oxide Nanoparticles—cont'd

Splits[a]						Quasi-SMILES for metal oxide nanoparticles	pLC_{50} in mol/L [30]
1	2	3	4	5	6		
v	t	c	t	t	c	O=[Zn]	5.80
c	v	t	c	t	v	O=[Si]=O^	2.92
v	c	c	t	v	v	O=[Zr]=O^	3.04
t	t	t	v	t	v	O=[Sn]=O^	3.24
t	t	t	t	t	t	O=[Ti]=O^	4.68
t	t	t	t	v	t	[Co]=O^	3.33
v	v	v	c	t	t	[Ni]=O^	3.87
t	t	t	t	t	t	O=[Cr]O[Cr]=O^	2.06
t	t	t	t	C	t	O=[La]O[La]=O^	5.56

SMILES, simplified molecular input-line entry system.
[a]*c, calibration set; t, training set; v, validation set.*

Table 8.2 contains the correlation weights for calculations with Eq. (8.13). An example of the calculation with Eq. (8.13) for quasi-SMILES is provided in Table 8.3.

8.3.2.4. Domain of Applicability

A value that characterizes half (50%) of any measured property is a widely prevalent measure for rationalization of the research work. Examples include the definition of bit (elementary quantity of information), lethal dose for half of the organisms (LD_{50}), the square of correlation coefficient that should be larger than 0.5 (i.e., again 50%), and so on. Inequality 12 gives the possibility of defining SMILES that fall into the domain of applicability of prevalence of different molecular features (extracted from SMILES). In addition, the percentage of SMILES that fall into the domain of applicability is a measure of quality for the split into the training and validation sets. One assumes that the split is satisfactory if more than 50% of the compounds are in the domain of applicability.

The percentages of the domain of applicability, according to inequality 12, are 76%, 76%, 76%, 71%, 71%, and 71%, for splits 1, 2, 3, 4, 5, and 6, respectively. As it was noted earlier, one can define 50% as a threshold to confirm acceptability of a split. Thus, a split that is characterized by the

TABLE 8.2 Correlation Weights for Calculations With Eq. (8.13)

x_{kj}	$CW(x_{kj})$	Frequency in Training Set	Frequency in Calibration Set
=	0.10158	22	6
Al	0.19940	1	0
Bi	1.00478	2	0
Co	1.35252	1	0
Cr	−0.24908	1	1
Cu	3.44583	2	0
Fe	0.20463	2	0
O	−0.27911	22	6
In	0.62103	1	0
La	1.95157	1	1
Ni	2.29898	1	1
V	0.79887	1	1
Sb	0.72663	1	1
Si	0.84577	2	0
Y	2.40333	1	0
Sn	1.05005	1	0
Ti	1.64851	2	0
[	0.20343	22	6
^	1.00105	12	2
Zn	4.60092	1	1
Zr	1.09949	1	0

domain of applicability of more than 50% can be considered as satisfactory: all six examined splits are satisfactory ones.

8.3.2.5. Mechanistic Interpretation

The obtained QSAR model allows evaluating the role of various eclectic features on the studied endpoint. Based on the developed model, one concludes that the double bonds (" = ") are stable promoters for the decrease of the cytotoxicity. The illumination represents a promoter of increase of the cytotoxicity for the metal nano oxides under study (Table 8.1).

TABLE 8.3 Example of Calculation of DCW(1,7) for Eq. (8.13)

x_{kj}	$CW(x_{kj})$
[	0.20343
Cu	3.44583
[	0.20343
=	0.10158
O	−0.27911
$\sum_{x_{kj} \in \text{quasi-SMILES}} CW(x_{kj})$	3.67516

The representation of metal oxide is "[Cu] = O".DCW(1,9) = $\sum CW(x_{kj})$ = 3.67,516. pLC50 = 1.5185 + 0.8370 * DCW(1,9) = 4.5946.

8.3.3. Membrane Damage by Means of TiO_2 Nanoparticles Under Different Conditions

8.3.3.1. The Description of Endpoint

A recent experimental study on membrane damage by metal oxide nanoparticles has provided interesting results that were used to develop another QSAR model. Experimental data on the physicochemical features of TiO_2 nanoparticles and their influence on membrane damage were taken from the literature (Sayes and Ivanov, 2010). These were (1) engineered size (nanometers), (2) size in water suspension (nanometers), (3) size in phosphate buffered saline (nanometers), (4) concentration (milligrams per liter), and (5) zeta potential (millivolts). Table 8.4 contains these parameters. The aforementioned physicochemical features of TiO_2 nanoparticles were involved in building up quasi-SMILES and QSAR models for membrane damage values related to various TiO_2 nanoparticles (characterized by different physicochemical features) (Toropova and Toropov, 2013). The physicochemical data were normalized using the following equation:

$$Norm(X_k) = \frac{\min X_k + X_k}{\min X_k + \max X_k} \tag{8.14}$$

Table 8.5 contains normalized data used to build up the quasi-SMILES. Table 8.6 contains quasi-SMILES defined according to the scale represented in Fig. 8.5. Three various splits of experimental data into the training and test sets were examined (Toropova and Toropov, 2013). These splits obey the following principles: (1) they are random and (2) the ranges of the endpoint for the training and test sets are similar.

TABLE 8.4 Experimental Data on Features (Impacts) of TiO_2 Nanoparticles and Their Denotations

	A	B	C	D	E
ID	Engineered Size, nm	Size in Water, nm	Size in PBS, nm	Concentration, mg/L	Zeta Potential, mV
1	30	125	1250	25	10
2	30	102	987	25	12
3	30	281	1543	50	15
4	30	101	1045	50	9
5	30	299	1754	100	11
6	30	134	961	100	11
7	30	600	1876	200	12
8	30	298	1165	200	12
9	45	129	2567	25	9
10	45	129	2309	25	10
11	45	201	2431	50	9
12	45	201	2987	50	11
13	45	451	2941	100	11
14	45	451	1934	100	9
15	45	876	1965	200	11
16	45	876	2109	200	10
17	125	136	3215	25	11
18	125	136	2667	25	10
19	125	149	3782	50	10
20	125	149	2144	50	15
21	125	343	3871	100	12
22	125	343	2890	100	9
23	125	967	3813	200	9
24	125	967	2671	200	8

TABLE 8.5 Normalized (Eq. 8.14) Representation of Physicochemical Features of TiO_2 Nanoparticles

	A	B	C	D	E
ID	Engineered Size, Normalized	Size in Water, Normalized	Size in PBS, Normalized	Concentration, Normalized	-Zeta Potential, Normalized
1	0.39	0.21	0.46	0.22	0.78
2	0.39	0.19	0.40	0.22	0.87
3	0.39	0.36	0.52	0.33	1.00
4	0.39	0.19	0.42	0.33	0.74
5	0.39	0.37	0.56	0.56	0.83
6	0.39	0.22	0.40	0.56	0.83
7	0.39	0.66	0.39	1.00	0.87
8	0.39	0.37	0.44	1.00	0.87
9	0.48	0.22	0.73	0.22	0.74
10	0.48	0.22	0.68	0.22	0.78
11	0.48	0.28	0.70	0.33	0.74
12	0.48	0.28	0.82	0.33	0.83
13	0.48	0.52	0.81	0.56	0.83
14	0.48	0.52	0.60	0.56	0.74
15	0.48	0.91	0.61	1.00	0.83
16	0.48	0.91	0.64	1.00	0.78
17	1.00	0.22	0.86	0.22	0.83
18	1.00	0.22	0.75	0.22	0.78
19	1.00	0.23	0.98	0.33	0.78
20	1.00	0.23	0.64	0.33	1.00
21	1.00	0.42	1.00	0.56	0.87
22	1.00	0.42	0.80	0.56	0.74
23	1.00	1.00	0.99	1.00	0.74
24	1.00	1.00	0.75	1.00	0.70

TABLE 8.6 **Building up Quasi-SMILES for Model of Membrane Damage Values by TiO_2 Nanoparticles (MD, Units/L)**

	A	B	C	D	E	
ID	Code for Engineered Size	Code for Size in Water	Code for Size in PBS	Code for Concentration	Code for Zeta Potential	MD, Units/L
1	A3	B2	C4	D2	E7	0.90
2	A3	B1	C4	D2	E8	1.00
3	A3	B3	C5	D3	E9	0.75
4	A3	B1	C4	D3	E7	0.70
5	A3	B3	C5	D5	E8	1.04
6	A3	B2	C3	D5	E8	1.09
7	A3	B6	C5	D9	E8	1.15
8	A3	B3	C4	D9	E8	1.20
9	A4	B2	C7	D2	E7	0.90
10	A4	B2	C6	D2	E7	0.85
11	A4	B2	C7	D3	E7	0.75
12	A4	B2	C8	D3	E8	0.78
13	A4	B5	C8	D5	E8	1.40
14	A4	B5	C5	D5	E7	1.50
15	A4	B9	C6	D9	E8	1.35
16	A4	B9	C6	D9	E7	1.40
17	A9	B2	C8	D2	E8	1.25
18	A9	B2	C7	D2	E7	1.17
19	A9	B2	C9	D3	E7	1.00
20	A9	B2	C6	D3	E9	1.10
21	A9	B4	C9	D5	E8	1.50
22	A9	B4	C7	D5	E7	1.42
23	A9	B9	C9	D9	E7	1.60
24	A9	B9	C7	D9	E6	1.65

SMILES, simplified molecular input-line entry system.

X=A,B,C,D,E

9, Norm(X)>0.9	**X9**
8, 0.8<Norm(X)<0.9	**X8**
7, 0.7<Norm(X)<0.8	**X7**
6, 0.6<Norm(X)<0.7	**X6**
5, 0.5<Norm(X)<0.6	**X5**
4, 0.4<Norm(X)<0.5	**X4**
3, 0.3<Norm(X)<0.4	**X3**
2, 0.2<Norm(X)<0.3	**X2**
1, 0.1<Norm(X)<0.2	**X1**
0, Norm(X)<0.1	**X0**

FIGURE 8.5 Partition of normalized physicochemical features into categories 1–9 according to its value.

8.3.3.2. The Description of Quasi-SMILES

Experimental data were used to develop quasi-SMILES for the investigated phenomena. Table 8.7 contains the correlation weights of various contributions used in the predictive model.

8.3.3.3. The Statistical Characteristics of Model

The best predictive model for membrane damage suggested in the work Toropova and Toropov, 2013 is the following:

$$\begin{aligned} &\mathrm{MD} = 0.8054\ (\pm 0.0044) + 0.1273\ (\pm 0.0014) \times \mathrm{DCW}(2,20) \\ &n = 10, r^2 = 0.9893, q^2 = 0.9845, s = 0.025, F = 741(\text{training set}) \\ &n = 5, r^2 = 0.9647, s = 0.066, (\text{calibration set}) \\ &n = 9, r^2 = 0.8679, s = 0.115(\text{validation set}) \end{aligned} \tag{8.15}$$

8.3.3.4. Domain of Applicability

The quality of the developed model was tested by investigation of the domain of applicability. All quasi-SMILES of the validation set fall into the domain of applicability according to inequality 12.

8.3.3.5. Mechanistic Interpretation

Based on the correlation weights obtained in three runs of the Monte Carlo optimization, one can conclude that A4 and A9 are promoters of increase of

TABLE 8.7 Correlation Weights (CWs) for Calculation of DCW(T^*,N^*)

Split 1		Split 2		Split 3	
x_{kj}	CW (x_{kj})	x_{kj}	CW (x_{kj})	x_{kj}	CW (x_{kj})
A3	−0.11150	A3	0.71150	A3	0.16450
A4	1.30300	A4	1.19800	A4	0.82400
A9	2.83850	A9	3.25100	A9	2.55400
B1	0.0	B1	0.0	B1	0.40000
B2	−0.69250	B2	−0.85300	B2	−0.04800
B3	0.15000	B3	−0.02800	B3	−0.05200
B4	0.0	B4	0.0	B4	0.0
B9	0.0	B5	1.88450	B5	2.42275
C3	0.0	B6	0.0	B9	2.41050
C4	0.0	B9	0.0	C3	0.0
C5	0.0	C3	0.0	C4	0.36575
C6	0.0	C4	0.0	C5	0.64075
C7	−0.74600	C5	−0.48000	C6	0.0
C8	0.0	C6	0.0	C7	0.58550
C9	0.0	C7	0.0	C8	−0.05400
D2	1.03450	C8	−0.04900	C9	0.63875
D3	−0.56350	C9	−0.20300	D2	1.08450
D5	2.67400	D2	1.03950	D3	−0.03000
D9	3.01650	D3	−0.21050	D5	2.63950
E7	0.14800	D5	2.34800	D9	2.48750
E8	0.27600	D9	3.10850	E6	0.0
E9	0.0	E7	0.56450	E7	0.10200
		E8	0.29250	E8	0.68875
		E9	0.0	E9	0.0

membrane damage caused by TiO_2 nanoparticles. On the other hand, B2 is the promoter of decrease for the endpoint. These findings help to shed some light on investigated phenomena. Again, the developed model allows to shed a light on the nature of the studied phenomena (Table 8.8).

TABLE 8.8 An Example of Model for TiO_2 Nanoparticles' Membrane Damage

Set	Quasi-SMILES	DCW *(2,20)*	Experimental	Calculated	Exprimental-Calculated	ID
Training	A3B3C5D3E9	−0.52500	0.750	0.739	0.011	3
Training	A3B2C3D5E8	2.14600	1.090	1.079	0.011	6
Training	A3B3C4D9E8	3.33100	1.200	1.229	0.029	8
Training	A4B2C7D2E7	1.04700	0.900	0.939	0.039	9
Training	A4B2C7D3E7	−0.55100	0.750	0.735	0.015	11
Training	A4B9C6D9E7	4.46750	1.400	.374	0.026	16
Training	A9B2C8D2E8	3.45650	1.250	.245	0.005	17
Training	A9B2C7D2E7	2.58250	1.170	.134	0.036	18
Training	A9B2C9D3E7	1.73050	1.000	.026	0.026	19
Training	A9B4C7D5E7	4.91450	1.420	1.431	0.011	22
Calibration	A3B1C4D2E8	1.19900	1.000	0.958	0.042	2
Calibration	A4B2C8D3E8	0.32300	0.780	0.847	0.067	12
Calibration	A9B2C6D3E9	1.58250	1.100	1.007	0.093	20

Calibration	A9B4C9D5E8	5.78850	1.500	1.542	0.042	21
Calibration	A9B9C9D9E7	6.00300	1.600	1.570	0.030	23
Validation	A3B2C4D2E7	0.37850	0.900	0.854	0.046	1
Validation	A3B1C4D3E7	−0.52700	0.700	0.738	−0.038	4
Validation	A3B3C5D5E8	2.98850	1.040	1.186	−0.146	5
Validation	A3B6C5D9E8	3.18100	1.150	1.210	−0.060	7
Validation	A4B2C6D2E7	1.79300	0.850	1.034	−0.184	10
Validation	A4B5C8D5E8	4.25300	1.400	1.347	0.053	13
Validation	A4B5C5D5E7	4.12500	1.500	1.331	0.169	14
Validation	A4B9C6D9E8	4.59550	1.350	1.390	−0.040	15
Validation	A9B9C7D9E6	5.10900	1.650	1.456	0.194	24

SMILES, simplified molecular input-line entry system.

8.3.4. Mutagenicity of Fullerene Under Different Conditions

8.3.4.1. The Description of Endpoints

Another study targeted the prediction of mutagenicity of the most classical nanoparticle—fullerene. The experimental study provided two endpoints. Both were examined in the computational work by Toropova et al., 2016a,b:

1. The bacterial reverse mutation test conducted using *Salmonella typhimurium* strains TA100 [in the presence and absence of metabolic activation under dark conditions and irradiation were taken from the work (Shinohara et al., 2009)] and
2. The bacterial reverse mutation test conducted using *Escherichia coli* strain WP2 uvrA/pKM101 [in the presence and absence of metabolic activation under dark condition and irradiation were taken from the literature (Shinohara et al., 2009)].

The experimental data allow considering a number of features that could be used to develop QSAR model. Twenty quasi-SMILES were defined for these data. These 20 quasi-SMILES were further randomly distributed into the training, calibration, and validation sets.

8.3.4.2. The Description of Quasi-SMILES

The details of the computational work are shown in Tables 8.9–8.11. Table 8.9 contains the scheme of building up quasi-SMILES. This provides the basis for the next steps of the study. Quasi-SMILES and experimental data on mutagenicity TA100 of fullerene under different conditions presented in Table 8.9 are displayed in Table 8.10. Table 8.11 contains the correlation weights used as the

TABLE 8.9 List of Conditions, Having Impact Upon Mutagenicity of Fullerene C_{60} Nanoparticles, Which Were Utilized to Build up Quasi-SMILES and Models

Conditions	Symbols for Quasi-SMILES
Dark or irradiation	"0" = darkness "1" = irradiation
Mix S9	"+" = with Mix S9 "-" = without Mix S9
Dose (g/plate)	"A" = 50 "B" = 100 "C" = 200 "D" = 400 "E" = 1000

SMILES, simplified molecular input-line entry system.

TABLE 8.10 Experimental and Calculated Values of the TA100 for Fullerene Nanoparticles Impact Under Different Conditions

ID	1[a]	2	3	Quasi-SMILES	Experiment	Split 1	Split 2	Split 3	1[a]	2	3
1	V	V	V	0 + A	146	132.8046	115.3775	143.2652	Y	Y	Y
2	T	C	V	0 + B	141	145.8861	121.3559	144.0029	N	Y	Y
3	T	C	C	0 + C	159	157.1709	136.5559	157.8729	N	Y	Y
4	V	C	V	0 + D	160	162.0685	142.5034	161.6878	Y	Y	Y
5	T	V	T	0 + E	177	165.3027	143.7145	165.5465	Y	N	Y
6	C	C	C	0 − A	143	130.9643	134.7925	147.8862	Y	Y	Y
7	T	T	C	0 − B	139	144.0458	140.7708	148.6238	N	Y	N
8	V	T	T	0 − C	169	155.3307	155.9708	162.4939	N	Y	Y
9	T	V	C	0 − D	168	160.2283	161.9183	166.3087	Y	Y	Y
10	T	T	T	0 − E	152	163.4625	163.1294	170.1675	Y	N	N
11	C	V	T	1 + A	129	116.4477	113.3044	130.1540	Y	Y	Y
12	T	C	T	1 + B	131	129.5292	119.2827	130.8917	N	Y	Y
13	V	T	T	1 + C	138	140.8141	134.4827	144.7618	N	Y	Y

Continued

TABLE 8.10 Experimental and Calculated Values of the TA100 for Fullerene Nanoparticles Impact Under Different Conditions—cont'd

ID	1[a]	2	3	Quasi-SMILES	Experiment	Split 1	Split 2	Split 3	1[a]	2	3
14	T	T	T	1 + D	137	145.7117	140.4303	148.5766	Y	Y	Y
15	C	V	T	1 + E	160	148.9459	141.6413	152.4354	Y	N	N
16	V	T	T	1 − A	136	114.6075	132.7193	134.7750	Y	Y	Y
17	T	T	V	1 − B	136	127.6890	138.6977	135.5127	N	Y	Y
18	T	T	C	1 − C	138	138.9739	153.8977	149.3827	N	Y	Y
19	C	T	T	1 − D	164	143.8715	159.8452	153.1976	Y	Y	Y
20	C	T	V	1 − E	172	147.1057	161.0562	157.0563	Y	N	N

SMILES, simplified molecular input-line entry system.
[a]*Split 1, 2, and three; C, calibration set; T, training set and V, validation set; Y, quasi-SMILES falls into Domain of applicability (otherwise "N").*

TABLE 8.11 Correlation Weights (CWs) for Symbols Which Represent Conditions of Fullerene Impact on TA100 Mutagenicity

Symbols of quasi-SMILES, x_{kj}	*CW (x_{kj}) run 1*	*CW (x_{kj}) run 2*	*CW (x_{kj}) run 3*
+	1.06698	0.47336	0.19086
-	0.99597	1.77789	0.56606
0	1.37861	0.96051	1.93566
1	0.74747	0.82121	0.87109
A	0.0	0.0	−0.05990
B	0.50476	0.40170	0.0
C	0.94020	1.42303	1.12619
D	1.12918	1.82266	1.43594
E	1.25398	1.90403	1.74925

SMILES, simplified molecular input-line entry system.

basis of the models for three different splits into the training, calibration, and validation sets. The quasi-SMILES and the experimental data on mutagenicity WP2uvrA/pKM101 of fullerene under different conditions (Table 8.9) are displayed in Table 8.12. They were used for model development, and Table 8.13 contains the correlation weights for the models considered in this study.

8.3.4.3. The Statistical Characteristics of Model

The utilization of the optimal descriptors calculated according to scheme suggested in the literature (Toropova et al., 2016a,b) resulted in the following best models for the two aforementioned endpoints:

$$\begin{aligned}
&\mathrm{TA100} = 117.813 + 12.3159 \times \mathrm{DCW}(2,3)\\
&n = 10, r^2 = 0.6810, s = 9.78, F = 17(\text{sub} - \text{training set})\\
&n = 5, r^2 = 0.9396, s = 7.91(\text{Calibration set})\\
&n = 5, r^2 = 0.7884, s = 7.79(\text{Validation set})
\end{aligned} \tag{8.16}$$

$$\begin{aligned}
&\mathrm{WP2uvrA/pKM101} = 84.9481 + 16.1111 \times \mathrm{DCW}(3,6)\\
&n = 10, r^2 = 0.6805, s = 12.1, F = 17(\text{sub} - \text{training set})\\
&n = 5, r^2 = 0.7480, s = 16.5(\text{calibration set})\\
&n = 5, r^2 = 0.8367, s = 25.7(\text{validation set})
\end{aligned} \tag{8.17}$$

TABLE 8.12 Experimental and Calculated Values of the WP2uvrA/pKM101 for Fullerene Nanoparticles Impact Under Different Conditions

ID	1[a]	2	3	Quasi-SMILES	Experiment	Split 1	Split 2	Split 3	1[a]	2	3
1	T	C	T	0 + A	113	118.9590	95.2449	133.2322	Y	Y	Y
2	T	V	C	0 + B	106	118.9590	127.8341	126.7746	Y	N	Y
3	V	T	V	0 + C	112	118.9590	127.9124	126.7746	Y	Y	Y
4	T	T	C	0 + D	115	118.9590	95.2449	126.7746	Y	Y	Y
5	T	C	T	0 + E	145	152.9472	169.4336	144.8871	Y	Y	Y
6	C	T	T	0 − A	160	159.2997	136.8401	162.2816	Y	Y	Y
7	V	T	C	0 − B	162	159.2997	169.4294	155.8240	Y	N	Y
8	C	V	C	0 − C	174	159.2997	169.5076	155.8240	Y	Y	Y
9	V	V	T	0 − D	179	159.2997	136.8401	155.8240	Y	Y	Y
10	T	T	V	0 − E	220	193.2879	211.0289	173.9365	Y	Y	Y
11	V	C	T	1 + A	114	84.2194	53.6913	104.3638	Y	Y	Y

12	C	V	V	1 + B	105	84.2194	86.2806	97.9062	Y	N	Y
13	V	C	V	1 + C	113	84.2194	86.3588	97.9062	Y	Y	Y
14	T	V	C	1 + D	110	84.2194	53.6913	97.9062	Y	Y	Y
15	C	T	T	1 + E	123	118.2076	127.8801	116.0187	Y	Y	Y
16	T	V	T	1 − A	127	124.5601	95.2866	133.4132	Y	Y	Y
17	C	T	T	1 − B	133	124.5601	127.8759	126.9556	Y	N	Y
18	T	T	V	1 − C	121	124.5601	127.9541	126.9556	Y	Y	Y
19	C	C	T	1 − D	117	124.5601	95.2866	126.9556	Y	Y	Y
20	T	C	T	1 − E	138	158.5483	169.4754	145.0681	Y	Y	Y

SMILES, simplified molecular input-line entry system.
[a]*Split 1, 2, and 3; T, training set; C, calibration set; V, validation set; Y, quasi-SMILES falls into domain of applicability (otherwise "N").*

TABLE 8.13 Correlation Weights (CWs) for Symbols Which Represent Conditions of Impact of Fullerene C_{60} Nanoparticles on Mutagenicity for WP2uvrA/pKM101

Symbols of quasi-SMILES, S_k	*CW (S_k) run 1*	*CW (S_k) run2*	*CW (S_k) run 3*
+	0.50000	0.69675	0.39982
-	0.94039	1.60115	2.20288
0	1.12778	1.60305	2.19630
1	0.74854	0.69727	0.40447
A	0.0	0.0	0.40082
B	0.0	0.99558	0.0
C	0.0	0.70300	0.0
D	0.0	0.0	0.0
E	0.37104	1.59979	1.12422

SMILES, simplified molecular input-line entry system.

8.3.4.4. Domain of Applicability

The domains of applicability for quasi-SMILES involved in building up models are presented in Table 8.10 (TA100) and Table 8.12 (WP2uvrA/pKM101).

8.3.4.5. Mechanistic Interpretation

Interestingly, almost all the correlation weights are positive for the mutagenicity models of fullerene TA100 and WP2uvrA/pKM101. However, their values are different. One can extract features of quasi-SMILES with relative large values. These features represent leading contributions to the investigated phenomena. The two largest contributions include darkness (0) and absence of Mix S9 (−). One needs to note that the obtained results are based on a small pool of experimental data. Apparently, it is possible that this interpretation can be adjusted after similar analysis is performed on larger experimental data for the studied endpoints.

8.4. CONCLUSIONS

The chapter reviews a concept of development of quasi-SMILES and their application to all existing experimental data available for the studied species. This is a major difference between traditional SMILES and quasi-SMILES approaches. The proposed concept has been used to predict outcomes of various processes involving nanomaterials. We do believe that the suggested hypothesis of building up predictive models is universal and can be relatively simply utilized to solve various nonstandard tasks. This allows extending applications of QSAR/QSPR techniques to the cases not covered by the traditional methods.

Unique abilities of nanomaterials are well known. The probability of these substances being effective pharmaceutical agents is high. However, traditional QSPR/QSAR analyses of these abilities (or these endpoints) are often not convenient for practice, whereas the quasi-SMILES described here give the possibility of solving tasks, which are unsolvable by traditional QSPR/QSAR paradigms.

ACKNOWLEDGMENTS

This work was financially supported by National Science Foundation: NSF-CREST Grant #HRD-1547754 and EPSCoR Grant #362492-190200-01\NSFEPS-0903787. A.A.T. and A.P.T. thank the EC project PeptiCAPS (Project reference: 686141).

REFERENCES

Bonchev, D., Balaban, A.T., Mekenyan, O., 1980. Generalization of the graph center concept, and derived topological centric indexes. J. Chem. Inf. Comput. Sci. 20, 106–113.

CORAL Software, http://www.insilico.eu/coral.

De Jong, W.H., Borm, P.J.A., 2008. Drug delivery and nanoparticles: applications and hazards. Int. J. Nanomedicine 3 (2), 133–149.

Duchowicz, P.R., Fioressi, S.E., Bacelo, D.E., Saavedra, L.M., Toropova, A.P., Toropov, A.A., 2015. QSPR studies on refractive indices of structurally heterogeneous polymers. Chemom. Intell. Lab. Syst. 140, 86–91.

Fourches, D., Pu, D., Tropsha, A., 2011. Exploring quantitative nanostructure–activity relationships (QNAR) modeling as a tool for predicting biological effects of manufactured nanoparticles. Comb. Chem. High Throughput Screen. 14, 217–225.

Glotzer, S.C., Solomon, M.J., 2007. Anisotropy of building blocks and their assembly into complex structures. Nat. Mater. 6, 557–562.

Gutman, I., Toropov, A.A., Toropova, A.P., 2005. The graph of atomic orbitals and it's basic properties. 1. Wiener index. MATCH Commun. Math. Comput. Chem. 53, 215–224.

Gutman, I., Furtula, B., Petrović, M., 2009. Terminal Wiener index. J. Math. Chem. 46 (2), 522–531.

Hosoya, H., 1972. Topological index as a sorting device for coding chemical structures. J. Chem. Doc. 12, 181–183.

Oksel, C., Ma, C.Y., Liu, J.J., Wilkins, T., Wang, X.Z., 2015. (Q)SAR modelling of nanomaterial toxicity: a critical review. Particuology 21, 1–19.

Pathakoti, K., Huang, M.-J., Watts, J.D., He, X., Huey-Min Hwang, H.-M., 2014. Using experimental data of *Escherichia coli* to develop a QSAR model for predicting the photo-induced cytotoxicity of metal oxide nanoparticles. J. Photochem. Photobiol. A: Chem. 130, 234–240.

Sayes, C., Ivanov, I., 2010. Comparative study of predictive computational models for nanoparticle induced cytotoxicity. Risk Anal. 30, 1723–1734.

Shinohara, N., Matsumoto, K., Endoh, S., Maru, J., Nakanishi, J., 2009. *In vitro* and *in vivo* genotoxicity tests on fullerene C_{60} nanoparticles. Toxicol. Lett. 191, 289–296.

Speck-Planche, A., Kleandrova, V.V., Luan, F., Cordeiro, M.N.D.S., 2011. Multi-target drug discovery in anti-cancer therapy: fragment-based approach toward the design of potent and versatile anti-prostate cancer agents. Bioorg. Med. Chem. 19 (21), 6239–6244.

Speck-Planche, A., Kleandrova, V.V., Luan, F., Cordeiro, M.N.D.S., 2012a. Chemoinformatics in anti-cancer chemotherapy: multi-target QSAR model for the in silico discovery of anti-breast cancer agents. Eur. J. Pharm. Sci. 47 (1), 273–279.

Speck-Planche, A., Kleandrova, V.V., Luan, F., Cordeiro, M.N.D.S., 2012b. Predicting multiple ecotoxicological profiles in agrochemical fungicides: a multi-species chemoinformatic approach. Ecotoxicol. Environ. Saf. 80, 308–313.

Speck-Planche, A., Kleandrova, V.V., Cordeiro, M.N.D.S., 2013. New insights toward the discovery of antibacterial agents: multi-tasking QSBER model for the simultaneous prediction of anti-tuberculosis activity and toxicological profiles of drugs. Eur. J. Pharm. Sci. 48 (4–5), 812–818.

Toropov, A.A., Toropova, A.P., Martyanov, S.E., Benfenati, E., Gini, G., Leszczynska, D., Leszczynski, J., 2011. Comparison of SMILES and molecular graphs as the representation of the molecular structure for QSAR analysis for mutagenic potential of polyaromatic amines. Chemom. Intell. Lab. Syst. 109, 94–100.

Toropov, A.A., Toropova, A.P., 2014. Optimal descriptor as a translator of eclectic data into endpoint prediction: mutagenicity of fullerene as a mathematical function of conditions. Chemosphere 104, 262–264.

Toropov, A.A., Toropova, A.P., 2015. Quasi-QSAR for mutagenic potential of multi-walled carbon-nanotubes. Chemosphere 124, 40–46.

Toropova, A.P., Toropov, A.A., Benfenati, E., Gini, G., Leszczynska, D., Leszczynski, J., 2011a. CORAL: quantitative structure–activity relationship models for estimating toxicity of organic compounds in rats. J. Comput. Chem. 32, 2727–2733.

Toropova, A.P., Toropova, A.A., Benfenati, E., Gini, G., 2011b. QSAR modelling toxicity toward rats of inorganic substances by means of CORAL. Cent. Euro. J. Chem. 9 (1), 75–85.

Toropova, A.P., Toropov, A.A., Benfenati, E., Gini, G., 2011c. Co-evolutions of correlations for QSAR of toxicity of organometallic and inorganic substances: an unexpected good prediction based on a model that seems untrustworthy. Chemom. Intell. Lab. Syst. 105, 215–219.

Toropova, A.P., Toropov, A.A., Rallo, R., Leszczynska, D., Leszczynski, J., 2012. Novel application of the CORAL software to model cytotoxicity of metal oxide nanoparticles to bacteria *Escherichia coli*. Chemosphere 89, 1098–1102.

Toropova, A.P., Toropov, A.A., 2013. Optimal descriptor as a translator of eclectic information into the prediction of membrane damage by means of various TiO_2 nanoparticles. Chemosphere 93, 2650–2655.

Toropova, A.P., Toropov, A.A., Benfenati, E., Puzyn, T., Leszczynska, D., Leszczynski, J., 2014. Optimal descriptor as a translator of eclectic information into the prediction of membrane damage: the case of a group of ZnO and TiO_2 nanoparticles. Ecotoxicol. Environ. Saf. 108, 203–209.

Toropova, A.P., Toropov, A.A., Rallo, R., Leszczynska, D., Leszczynski, J., 2015. Optimal descriptor as a translator of eclectic data into prediction of cytotoxicity for metal oxide nanoparticles under different conditions. Ecotoxicol. Environ. Saf. 112, 39–45.

Toropova, A.P., Toropov, A.A., Veselinović, A.M., Veselinović, J.B., Benfenati, E., Leszczynska, D., Leszczynski, J., 2016a. Nano-QSAR: model of mutagenicity of fullerene as a mathematical function of different conditions. Ecotoxicol. Environ. Saf. 124, 32–36.

Toropova, A.P., Achary, P.G.R., Toropov, A.A., 2016b. Quasi-SMILES for Nano-QSAR prediction of toxic effect of Al_2O_3 nanoparticles. J. Nanotoxicol. Nanomed. 1 (1), 17–28.

Webster, D.M., Sundaram, P., Byrne, M.E., 2013. Injectable nanomaterials for drug delivery: carriers, targeting moieties, and therapeutics. Eur. J. Pharm. Biopharm. 84, 1–20.

Weininger, D., 1988. SMILES, a chemical language and information system. 1. Introduction to methodology and encoding rules. J. Chem. Inf. Comput. Sci. 28 (1), 31–36.

Weininger, D., Weininger, A., Weininger, J.L., 1989. SMILES. 2. Algorithm for generation of unique SMILES notation. J. Chem. Inf. Comput. Sci. 29 (2), 97–101.

Weininger, D., 1990. SMILES. 3. Depict. Graphical depiction of chemical structures. J. Chem. Inf. Comput. Sci. 30 (3), 237–243.

Wiener, H., 1947a. Correlation of heats of isomerization, and differences in heats of vaporization of isomers, among the paraffin hydrocarbons. J. Am. Chem. Soc. 69 (11), 2636–2638.

Wiener, H., 1947b. Structural determination of paraffin boiling points. J. Am. Chem. Soc. 69 (1), 17–20.

Wiener, H., 1948. Relation of the physical properties of the isomeric alkanes to molecular structure. surface tension, specific dispersion, and critical solution temperature in aniline. J. Phys. Chem. 52 (6), 1082–1089.

Index

'*Note*: Page numbers followed by "f" indicate figures, "t" indicate tables.'

Q

R

S

T

Made in the USA
Las Vegas, NV
28 October 2024

10605107R10143